SEVEN ELEMENTS
THAT CHANGED THE WORLD

'The human quest for knowledge and insight has led to extraordinary progress. It has transformed the lives we lead and the world we live in. But that onward march has also thrown us huge challenges about how we treat each other and the planet on which we live. This book forces us to confront these realities and does it in a unique and fascinating way. It weaves science and humanity together in a way that gives us new insight. This is an expertly crafted book by a unique thinker and talented engineer and businessman.'
– **Tony Blair**

'The progress and prosperity that humanity has achieved', writes John Browne, 'is driven by people – scientists, business people and politicians.' The author has the rare distinction of having wide and deep experience of all three fields, and this is what makes *Seven Elements* such a fascinating and enjoyable book. Part popular science, part history, part memoir, these pages are infused with insight, shaped by the experience of a FTSE 100 Chief Executive and lifted by the innate optimism of a scientist.'
– **Brian Cox**

'*Seven Elements* is a boon for those, like me, who gave up science much too soon in our teens. John Browne has found a fascinating way of helping us break through the crust of our ignorance. The scientific literate too will relish his personal mix of historical knowledge and technical prowess with his gift for making the complicated understandable.'
– **Peter Hennessy**

'John Browne uses seven elements, building blocks of the physical world, to explore a multitude of worlds beyond. From the rise of civilizations, to some of today's most important challenges and opportunities, to the frontiers of research, he weaves together science, history, politics and personal experience. Browne tells a lively story that enables us to see the essential elements of modern life in a new, original and highly engaging way.'
– **Daniel Yergin,** Pulitzer Prize-winning author of *The Quest: Energy, Security and the Making of the Modern World* **and** *The Prize*

Also by John Browne

Beyond Business

SEVEN ELEMENTS
THAT CHANGED THE WORLD

SEVEN ELEMENTS
THAT CHANGED THE WORLD

AN ADVENTURE OF INGENUITY AND DISCOVERY

JOHN BROWNE

PEGASUS BOOKS
NEW YORK LONDON

SEVEN ELEMENTS THAT CHANGED THE WORLD

Pegasus Books LLC
80 Broad Street, 5th Floor
New York, NY 10004

ISBN: 978-1-60598-540-4

10 9 8 7 6 5 4 3 2 1

Printed in the United States of America
Distributed by W. W. Norton & Company

To QNN

CONTENTS

PREFACE

WHY SEVEN?

The number seven has always held a central place in myth, music and literature. The world was created in seven days; there are seven notes in the diatonic scale; and, according to Shakespeare, there are seven ages of man. In conceiving this book, I was also drawn to the number seven and so I asked myself: which of the seven chemical elements help us best to understand our world and how it came to be? I also thought about which have had the greatest influence on my life and which I have experienced most directly.

Carbon, which in combination with hydrogen forms the bulk of crude oil, was obvious. So too was iron, the backbone of all industry since the eighteenth-century Industrial Revolution (and without which no oil could be extracted). Silver came next to my mind as the element that made possible photography, one of my lifelong passions. Looking for further inspiration, I found in my library my school copy of the periodic table, which organises the elements according to their chemical properties. As I scanned the chequerboard, from left to right, I passed along the elements, each containing one more proton in its nucleus than the last.[1]

First is hydrogen, vitally important in combination with so many other elements to form the structures of life and, as a result, fossil fuels.[2] But in its own right, hydrogen did not seem world-changing. Passing further along, I came to silicon, sitting directly below carbon as both elements contain four electrons in their outermost shell. I thought back to my time on the board of Intel, the pioneers of the silicon microchip. Their ubiquitous nature in our day-to-day lives – in making possible our digital world

– made silicon another obvious choice for inclusion.

Appearing in its world-changing form at the same time as silicon in the 1940s was titanium, the next element I stopped at. Once it was going to be the miracle element, a dream that did not quite work out. But what most drew me to it was its little-known use as a whitening agent in almost everything that is white. I learnt of this through business with Quebec Iron and Titanium in Canada. It surprised me then, and continues to astound me now.

Traversing the remainder of this line, I passed a number of familiar metals: iron, cobalt, nickel, copper, zinc. All of them are so important but, I wondered, which one actually changed the world. I stuck with my choice of iron and left copper behind; electrical engineering will have its fair share with silicon.

I passed silver, the element of photography, and then in the line below reached gold. Its universal allure led to its use in coins, the basis of currencies for centuries and the foundation of international trade. Gold became a great motivator for global expansion and imperial ambition. But the same attraction has led many to commits acts of immense cruelty. It continues to captivate us today.

Finally, I reached the bottom of the periodic table, six elements in tow. Here I came to uranium, whose nucleus, having accumulated so many protons and neutrons on the journey down the periodic table, is very unstable. That characteristic defined the post-war era on a day in 1945 in a city in Japan, and for that reason it is the seventh element.

Time and again, while writing this book, I have revisited the periodic table, questioning this choice of seven elements and questioning the choice of the number seven. Iron, carbon, gold, silver, uranium, titanium, silicon; each time, *these* seven elements have stood out as having most powerfully changed the course of human history. These seven elements have shaped the vast complexities of our social, economic and cultural existence. These seven elements hold a grip on our emotions – and our history – like no others.

I cannot think of an eighth.

The Essence of Everything

The elements are the source of all human prosperity and a great deal of human suffering. In numerous ways, I have seen both. Over the course of my forty-five years in business, including twelve as the leader of BP, I saw the very best and the very worst that the elements can do for humanity.

As a child, when I asked my father to tell me a story, improbably he really did begin with 'once upon a time ...'. That is where the story of the elements begins. If you pointed a very powerful radio telescope out into the sky, you would detect a stream of low-energy radiation coming from every direction. This radiation has been travelling undisturbed through space ever since the first elements were formed some fourteen billion years ago. It is the remnant, or echo, of the Big Bang that gave birth to the Universe.

At first, the Universe was nothing more than a fluid of pure energy. As it expanded and so cooled, particles, which are the basic building blocks of matter – protons, neutrons and electrons – appeared from the fluid. The Universe kept cooling and allowed the particles to fuse together to become helium and deuterium (a heavy form of hydrogen). This process of nuclear fusion would later give birth to all the other elements inside the stars.

I would ask my father to tell me stories about science, but he would not because he did not like the subject. To keep me quiet, he gave me a book of Christmas lectures by the physicist Sir William Bragg, originally delivered at the Royal Institution in 1923. In *Concerning the*

Nature of Things, Bragg describes how atoms of different elements could join to form the vast complexity of the world around us.[1] At some stage, they had then combined to create life itself with its astounding ability to shape our chaotic world. I was amazed that, at a fundamental level, our own lives and even our thoughts are simply the result of these atomic interactions. In the early twentieth century, Bragg and his son Lawrence were pioneers in the field of x-ray crystallography. They used x-rays to look at matter in unprecedented detail.[2] With these 'new eyes', the Braggs transformed our view of the elements, just as John Dalton's atomic theory and Mendeleev's periodic table had done in the century before.[3]

As a teenager growing up in southern Iran, where my father was stationed, I was surrounded by oil and its awe-inspiring industry. I was thrilled by watching the huge machinery which drilled the wells that produced the oil. As I had learnt from Bragg's lectures, oil is composed of hydrogen and carbon. 'Under the proper stimulus and in the presence of oxygen,' wrote Bragg, 'the atoms rush into fresh combination, developing great heat in doing so.'[4] I was fascinated by the process of transformation which produced the energy to transform society. Carbon, in the form of hydrocarbons, brought people heat, light and mobility, and so created freedom and new ways of life.

Nowhere was that more evident than in China. On my first visit in 1979, only three years after the death of Mao, the country was poor, bleak and bland. There were hardly any motor vehicles on the streets, merely a monochrome sea of miserable men and women in grey-green suits travelling by foot or by bicycle. Today China feels like the centre of the world, overflowing with skyscrapers and cars and bustling with people. Hundreds of millions of people have found prosperity in this transformation, a transformation that has been fuelled by carbon-based energy, of which China is now the world's largest consumer.[5]

In Azerbaijan, at the other end of Asia, I saw how hydrocarbons could bring great benefits to a country. The most visible beneficiaries appeared to be the ruling elite associated with allegations of corruption and the abuse of power, but there were real economic benefits to its citizens. The oil pipeline from Baku, the capital of Azerbaijan on the Caspian

Sea, to Ceyhan on the shore of the Turkish Mediterranean, completed in 2005, stretches for a thousand miles, through three countries and the lands of more than a hundred ethnic groups. More than 30,000 contracts were signed to secure the rights of the local people. As a result, the pipeline and the oil that flows through it have provided many benefits to the people of Azerbaijan, tripling the average income over the last decade.[6]

China and Azerbaijan are just two examples of how hydrocarbons, our primary fuel source since the Industrial Revolution, can transform our way of life for the better. But there and elsewhere, I saw carbon bring pollution and pain alongside prosperity.

En route to Anchorage, Alaska, in 1989, I looked out of the window of the plane to see below us the *Exxon Valdez* that had earlier run aground. Oil was flowing out of her side, coating the water and the white ice in satin black. It was an extraordinarily powerful image, which remains with me to this day, of the harmful impact that hydrocarbons can have on the natural world.

Elsewhere, greed, fuelled by carbon, has caused more than physical hurt to people and the environment; it has changed people's very nature, bringing out their darkest side. In the 1990s, I was responsible for a huge Colombian oilfield, located in the foothills of the Llanos Mountains in an area rife with drug lords, paramilitaries and bandits who were drawn to the oil like flies to a carcass. To protect ourselves we built a tall barbed-wire fence and surrounded ourselves with armed guards. People outside the fence soon grew to despise us and kidnappings and attacks became frighteningly common. They saw us profiting from a natural resource that they believed belonged to them, and they wanted a share of the returns to remain in their community. We responded by building taller fences, travelling everywhere by helicopter and bringing in the Colombian army. All sides were overcome with fear, anger and greed, fuelling human division, hatred and ultimately war.[7]

But carbon is not my only focus. Of the ninety-eight naturally occurring chemical elements on the periodic table, there are six others that have most powerfully changed the course of human history: iron, gold, silver, uranium, titanium and silicon. This book traces the story of how they have enabled progress as well as destruction, of the power they give humans to do good and evil, and of their capacity to shape our future.

Progress

For the greater part of our existence, we lived more like lower-order animals than humans, spending our days on the most basic of activities, searching for food, water and shelter. In that existence, there was no choice: everything was done to survive. About 50,000 years ago, humanity took a 'Great Leap Forward' with a wave of behavioural innovations that included the start of complex language, the first cave art, the origins of religious ritual and the beginnings of barter trade.[8] The timing and the origins of these changes are disputed, but there is little doubt that it was intimately connected to our use of the elements; new ways of carving limestone, creating iron pigments, and controlling wood fires. The creative use of the elements made our survival less burdensome and gave humanity the tools to lay the foundations of civilisation. Beyond this, they have continued to give us the means to do great things, to give us more freedom and to give us more choices in the conduct of our daily lives.

Human progress can be measured by our ability to harness greater amounts of energy and so transform the world far beyond what would be achievable by human strength alone. No energy source has been more potent than carbon, in the form of wood, coal, oil and natural gas. Coal enabled industrial revolutions in Europe and the US, as it gave us the ability to expand our productivity; an amount of coal equal to the weight of an average man can do the same work as that man working for a hundred days. We have used carbon to accomplish extraordinary feats in our endeavours, whether in travel, trade, art, engineering or communication.

Carbon, too, has unlocked the potential of the other elements: with its energy, we have smelted iron, mined gold and enriched uranium. It is the creative force that underpins all others. Carbon's most powerful alliance is with iron. We need only to look around, at the railways, the factories and the skyscrapers, to see how the wealth of industry and the fabric of society are built from iron.

In the most specialist applications, for which iron is too weak or heavy, futuristic titanium metal has been used to accomplish triumphs of air and sea exploration. But far more pervasive than titanium's use as a metal in supersonic aircraft and deep-diving submarines is its use as bright white titanium dioxide. In that form, titanium is everywhere around us, feeding our obsession with purity, cleanliness and façade. Milk is no purer and shirts are no cleaner as a result of the titanium dioxide that whitens them. It is their whiteness that satisfies some urge within us.

However ubiquitous, we do not normally notice titanium's presence in our regular lives. The same is true of silver in its use in photography. The impact of photography is so significant since it has enabled us to see the world in a way that we would not have otherwise been able to do. It has shown us the vivid reality of the Second World War, the Vietnam War and the Rwandan genocide. It has impacted the way we think about each other, by putting a human face to our leaders, our neighbours and our enemies. Perhaps most powerfully of all, silver has changed the way we think about ourselves. It records our memories, our histories and our relationships not as words or thoughts, but as lasting images.

Silver is much better known, along with gold, as a store of value and medium of trade. Ever since the first coins were minted over two millennia ago, possibly in the ancient city of Sardis, merchants have relied on the standards established by these rare and precious metals for international commerce. Gold and silver have enabled the movement of people and materials and the cross-fertilisation of ideas. They have not only helped to spread the economic benefits of the Earth's elements across the world but have also stimulated human progress.

Silicon is the final element of the story, and perhaps the most transformative of all. It was first used to make objects of beauty in the form of glass beads, vases and mirrors. Later it became a common, utilitarian building material, draped around the outside of skyscrapers, satisfying the human desire for light. But silicon's greatest impact has been in the last half-century as the inner workings of computers. In this 'Silicon Age', we calculate and communicate effortlessly, with instant access to the sum of human knowledge. Silicon's impact on society is perhaps greatest when placed in the hands of the ordinary citizen. As the heart of modern communication, silicon has supported political revolutions in the Arab Spring and broken down the geographical barriers that have restrained our social interactions for millennia.

Destruction

The elements have created progress, innovation and prosperity, but they have also wreaked great destruction on people and nature. Carbon's destructive force is felt through the indirect consequences of its extraction and consumption. During the Industrial Revolution in Great Britain, the air became thick with smoke and thousands died in mine collapses and explosions. As industrial revolutions followed around the world, the consequences were similar. Only in the last two decades have we come to realise carbon's most insidious effect. Burning hydrocarbons has released billions of tonnes of carbon dioxide into our atmosphere, trapping the energy of the sun and potentially changing the world's climate.

Often the destructive forces of the elements are unleashed by deliberate human action. The strength of iron has made it not only the beneficial tool of peaceful industry, but also the brutally efficient and bloody weapon of war, in swords, guns, ships and tanks. Iron has also been the subject of conflict: for almost a century, the great powers of Europe went to war to obtain control of the vast iron ore and coke reserves of the Ruhr and Alsace Lorraine.

Throughout my career I have seen how oil, the 'black gold', has driven men's passions, desires and greed. The world has become very dependent on oil and therefore anxious about securing reliable supplies of it. Oil confers powers on leaders who control it but is sometimes more of a curse than a benefit to the countries that produce it.

But in the history of the elements, humanity has committed the greatest acts of cruelty in its quest for ownership of gold. Over half a millennium, this precious metal has inspired intense greed, madness and violence, driving people to plunder, kill and enslave.

One element stands above all others in its destructive power. Uranium is the element which defined the post-war era. It is tied to one of the darkest moments in human history: the detonation of an atomic bomb over Hiroshima. From that dark moment came the great hope that we could use uranium's extraordinary energy for creation rather than destruction. But the great hope of cheap and abundant nuclear-generated electricity has been dogged by dread and fear. Uranium continues to command power on the global stage as we struggle to control the spread of nuclear weapons. By unlocking its power, we have created the potential for our own destruction.

Human choice

So great is the influence of these elements that they have taken on personalities of their own: uranium, the powerful and the fearful; gold, the alluring and hypnotic; and iron, the strong and dependable. But, in a sense, their story is nothing more than the story of seven arrangements of protons, neutrons and electrons, the pattern which gives each element its character. It is tempting to think of these characteristics as inevitable or even uncontrollable. But each element's character is determined by the choices we make. We are in control of our own destiny, and the elements are merely the tools for our progress or our destruction. We are not slaves of the elements; we are their masters.

And so this book is not about the elements *per se*. Rather, it is about how people have harnessed the intrinsic powers of the elements to shape our cultural, economic and social existence, and in doing so have transformed our world. I have seen much of this transformation first-hand, and so this story of seven elements also contains a personal element. It takes you on a journey of my adventures with oil barons in Russia, merchants in Venice, tribesmen in Colombia and computer wizards in Silicon Valley. And along the way, we explore the stories of remarkable times and remarkable individuals – Pizarro, Rockefeller, Carnegie, Curie – and their deep connection with the elements. They changed the course of history. They demonstrated the elements' latent potential to inspire and equip good men to do good and evil men to do evil. Whether we continue to use these elements for common human progress and prosperity, or for individual greed and iniquity, is up to us.

The American physicist Richard Feynman summed it up through a Buddhist proverb: 'To every man is given the key to the gates of heaven; the same key opens the gates of hell.'[9]

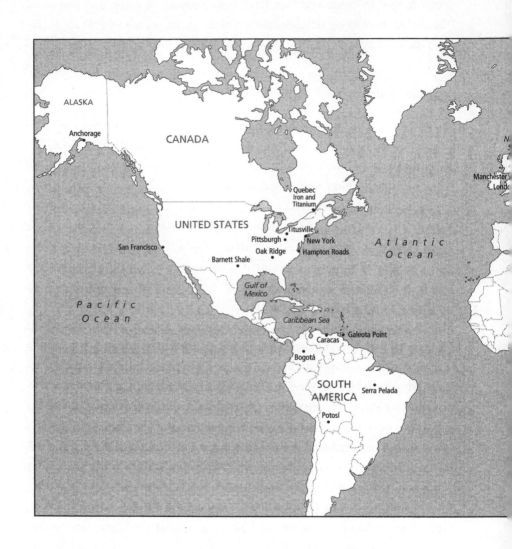

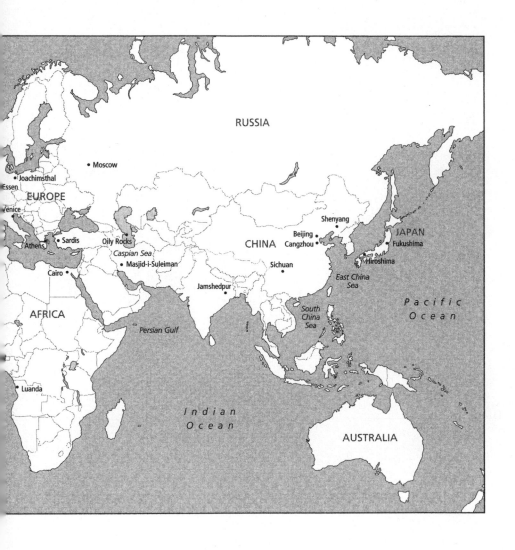

IRON

Battle of the ironclads

Ploughing with ease into her opponent's wooden frame, the ironclad Confederate State Ship *Virginia* marked a turning point in naval history. 'The crash below the water was distinctly heard,' recalled the flag officer of the opposing USS *Cumberland*, 'she commenced sinking, gallantly firing her guns as long as they were above water.'[1] But her fire simply bounced off the *Virginia*'s impenetrable iron hull.

During the American Civil War in March 1862, the CSS *Virginia* attacked the Federal ships at Hampton Roads in Virginia. The sinking of USS *Cumberland* led to the loss of about a third of its crew in what an officer on her deck described as 'a scene of carnage unparalleled in the water'.[2] The *Virginia* had been rebuilt from a sunken wooden-framed ship, the USS *Merrimack*, with makeshift equipment and poor engines. She had one great advantage: her two-inch-thick armoured plate which her opponent's wooden ships were unable to break. The Union forces panicked; if the *Virginia* could overcome the Union blockade at Hampton Roads, she could steam up the Potomac and shell Washington. That evening President Lincoln 'went repeatedly to the window and looked down the Potomac – the view being uninterrupted for forty miles – to see if the *Merrimack* was not coming to Washington'.[3]

Fortuitously, the Union forces had been developing their own ironclad, the *Monitor*, with an even thicker plate of eleven inches. Hearing of the advance of the *Virginia*, the ship set sail for Hampton Roads. The next day the first ever clash between ironclads took place. A lithograph depicting the conflict, made by the prolific printmaking firm of Currier & Ives,

1

hangs in my office.[4] I bought the print a long time ago because I liked the battle scene, not realising its significance. In the foreground the smaller and lighter *Monitor* darts towards the *Virginia*, both ships with guns blazing, smoke and steam billowing from their decks.[5] 'No battle that was ever fought, caused as great a sensation throughout the civilized world,' wrote eyewitness naval officer William Harwar Parker.[6]

It was an arduous fight: the ships engaged for more than four hours at close range. At first the *Virginia* fired exploding shells and the *Monitor* flung back solid shot, but both simply bounced off the iron hulls 'with no more effect, apparently, than so many pebbles stones thrown by a child'.[7] Soon they resorted to ramming tactics, but, by mid-afternoon, with no fatalities, the two vessels disengaged. The ships suffered only dents, and the crews, sealed in isolation behind thick iron walls, were virtually unhurt.[8] Sitting down to eat after the battle, the crew of the USS *Monitor* were all in high spirits. 'Well gentlemen,' said Assistant Secretary Gustavus Fox, coming on board later to commend the crew, 'you don't look as though you were just through one of the greatest naval conflicts on record.'[9]

Iron had embodied masculine strength and aggression long before the Battle of Hampton Roads. Its strength is one of the reasons why life is possible on this planet. Most of the Earth's core is made of iron. As the solid inner core spins, and conversion currents surge through the liquid outer core, a magnetic field is produced around the Earth. This keeps at bay the solar wind, an ionising radiation harmful to life. The first human uses of iron are difficult to trace due to the ease with which the metal corrodes, meaning that ancient iron objects are much rarer than those made of more durable metals such as gold and silver.[10] However, iron objects begin to appear after approximately 3500 BC in the form of jewellery, domestic implements and, most importantly, weapons. Iron went on to be used as a bloody tool of ancient war in the form of iron swords, shields and spears.

But for thousands of years warships were still built out of fragile and flammable wood. In the background of the Currier & Ives lithograph, these wooden ships keep their distance, an outclassed and soon to be outdated instrument of war. The Battle of Hampton Roads was proof to the tens of thousands of troops, watching from the estuary banks, of the superior might of the ironclad. At the beginning of America's Industrial Age, the

Virginia and the *Monitor* were the realisation of the power of industrial iron armoury, a force which would go on to shape the politics and wars of the modern world.

The element of peace

Across the Atlantic, in Germany, the 1860s were the start of an era of great industrial progress and prosperity. The Industrial Revolution had swept out of Great Britain and across Europe. Sitting on the banks of the River Ruhr, the city of Essen was the industrial centre of Germany. Small hillside blast furnaces had been replaced by colossal industrial factories and the once medieval market town was expanding quickly. During the decade, Essen's population rose by 150 per cent. One family, above all others, was responsible for this growth.

In 1587, Arndt Krupp joined the merchants' guild of Essen. He was the founder of the Krupp dynasty that would last for nearly four hundred years and become the leader not only of Germany's industrial prowess, but also of its machinery of war.

In their armament factories, Alfred Krupp, a descendant, forged the cannons for the wars led by Otto von Bismarck against Austria and France in 1866 and 1870. These weapons were decisive. The cast-iron artillery cannons of the Prussian army had twice the range and were far more accurate and more numerous than the French bronze pieces. In 1862, Bismarck famously declared that the German Empire would not be built on 'speeches and majority decisions' but on 'blood and iron'.[11] Whoever mastered iron, he believed, would master Europe.

In both world wars, Krupp's armaments again proved critical. The vast arsenal of the German army underpinned her strategic campaigns against the enemy. At the start of the First World War, long-range Krupp cannons smashed Belgian forts on their way towards Paris. In the Second World War, Krupp siege guns would fire shells weighing seven tonnes a distance of up to 40 kilometres.[12] The Krupp's iron forges supplied the munitions that enabled Germany to make war. But wars were not only fought with iron, they were also fought over reserves of iron ore and coke. Smelting iron ore with coke produces iron and carbon dioxide. During the

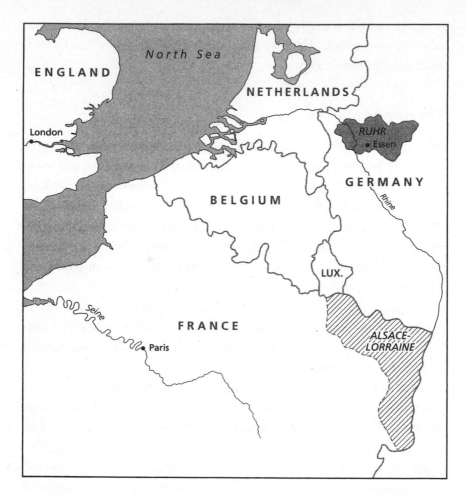

industrial revolutions, securing these reserves became a preoccupation among European nations. No one wanted to fall behind in this period of unprecedented economic growth.

The Ruhr region, in which the Krupp dynasty thrived, contained vast reserves of coal and somewhat smaller reserves of iron ore. In the late nineteenth and early twentieth centuries, these reserves became a source of great conflict, during which time France and Germany went to war three times.

In July 1870, France declared war on neighbouring Prussia. Prussia, with the bordering German Confederation states, which it often led, had become an increasing threat over the previous decade. Only four years earlier, Prussia had invaded Austria, leading to the creation of the powerful North German Confederation. France's once small and manageable neighbour now had both a formidable army and a flanking position on her border. Prussia's population was growing rapidly and its heavy industries were becoming dominant. By 1867, coal mines in Prussia and Saxony (another member of the North German Confederation) were outproducing French mines by three to one. France was being squeezed and decided to go to war.

But France underestimated just how strong Prussia had become. In a matter of weeks, the Prussian army advanced to Paris. After a siege lasting several months, the city fell on 28 January 1871 and the war ended. Prussia had destroyed France's military power and, as a requirement of the Treaty of Frankfurt, it was required to cede German-speaking Alsace Lorraine, which held valuable iron ore reserves. Only forty years later, France would fight against the now unified German Empire in the First World War. It would regain Alsace Lorraine, once again taking control of the region's iron ore reserves. France was now able to increase its production of steel but as a result it became even more dependent on the coke and coal needed for smelting.[13] When Germany defaulted on war reparation payments, France retaliated by invading the Ruhr. This not only secured coal supplies, but also crippled Germany's own industries. In response, Hitler began to remilitarise the Rhineland, in which the Ruhr sits. Wanting to avoid another war, France put up little resistance, giving Hitler the confidence to pursue a series of increasingly aggressive actions that ultimately led to the Second World War.[14]

The Ruhr's coke reserves were indispensable in the development of Europe's iron and steel industries. But the same resources made it a battleground for almost eighty years. During this time, the Ruhr rose to become the industrial heartland of Europe, but the region's success was also its downfall. In March 1943, the Allied forces made the first of what would become two hundred major air raids on Essen. More than 36,000 tonnes of incendiary and explosive bombs were dropped, the greater part of which landed on the eight square kilometres of Krupp factories. After the war, Essen became a bleak and cratered wasteland.[15] But in little more than five years the Ruhr would be rebuilt and integrated into a new political system that was designed to make iron a tool for peace rather than war.

On 9 May 1950, France's Foreign Minister Robert Schuman made a historic announcement on the radio: France was ready to partner with Germany, and other nations, to form a new European heavy industry community. The European Coal and Steel Community (ECSC) was founded in the aftermath of the Second World War in the hope of ending decades of economic and military competition. By pooling coal and steel resources, Schuman hoped to create a common foundation for economic development which he believed would make war 'not merely unthinkable, but materially impossible'.[16] Regions that had 'long been devoted to the manufacture of munitions of war, of which they had been the most constant victims' would now use iron to drive industrial development and raise living standards.[17] Schuman believed his simple yet bold plan would herald a new age of growth and prosperity.

The ECSC was the first step in the formation of the European Union, whose twenty-seven member states now constitute the largest economy in the world.[18] It was Europe's first major experiment in supranationalism, forming the foundation of a new entity which was both more stable and integrated. In return for sacrificing a degree of national sovereignty, members would reap economic and political benefits, not least the promise of peace.[19]

The impact is apparent today in the surrounding areas of Essen, which has been transformed from the 'forge' into the 'desk' of the Ruhr. It is a comfortable, modern city, home to many of Germany's largest corporations, not least of Aral, the face of BP in Germany. The Krupp family line had ended, but the name remains in the multinational conglomerate ThyssenKrupp. The Krupp Belt, once overflowing with industrial factories, is now home to the company's modern headquarters and what remains of the region's industrial past is now nothing more than a museum piece.[20]

The unity of the European Union and its predecessor entities has sustained an unprecedented period of peace.[21] Nations have been bound together through not only the interdependence of trade but also the security of common laws. It all started with carbon, as found in coal, and steel, as made from iron. It has been such a powerful tool for securing peace and prosperity because of the great extent to which these elements underpin modern society. Iron is everywhere, used in the construction of monumental skyscrapers, aeroplanes and wind turbines.[22] And, for me, one colossus stands above all others as a symbol of steel's might and an extraordinary example of what humanity has achieved with iron.

Sixty thousand tonnes of steel

11 July 2005: it was the ninetieth anniversary of BP Shipping, and to celebrate we held a party at the National Maritime Museum in Greenwich, south-east London. Cocktails were served under the arches of the museum and guests ambled around the 'Nelson & Napoleon' exhibition, marking two hundred years since the Battle of Trafalgar. We sat down to dinner, under the glass dome of Neptune Court, and Bob Malone, the leader of BP Shipping, stood up to give a speech. A lot had happened in those ninety years: for example, BP had at one time owned and operated the world's largest merchant fleet which, during the Second World War, had provided a good part of the transportation of fuel to the Allied forces.[23]

Bob finished speaking and we stood to toast the health of the company, but my thoughts were elsewhere. In the car on the way to the dinner that evening I had received a deeply worrying phone call from Tony Hayward, then leading BP's exploration and production activities. 'It's *Thunder Horse*,' he said, referring to our pioneering offshore oil platform in the Gulf of Mexico. 'It seems to be sinking.'

Thunder Horse PDQ is the biggest semi-submersible offshore production platform in the world, 50 per cent bigger than the previous record holder in Norway.[24] The hull alone is a 60,000-tonne mass of thick steel plate that holds a complex network of 50 kilometres of pipe work and 250 kilometres of cabling. This unprecedented construction was necessary to tap the Thunder Horse field, the biggest in the Gulf of Mexico with an expected production of a quarter of a million barrels of oil and 5.6 million cubic metres of natural gas each day. Only strong, abundant and cheap steel could be employed on such a scale and in such hostile marine conditions.

No existing vessel was big enough to transport *Thunder Horse*'s hull from the construction yard at Okpo, South Korea, to the Gulf of Mexico. The MV *Blue Marlin*, at the time one of the two biggest heavy-lift barges in the world, had to be modified by widening its hull and adding a new propulsion system. Even with the modifications, *Thunder Horse* still overhung the ship by 20 metres on either side. Too wide to fit through the Panama Canal and too tall to fit under the Suez Canal Bridge, *Thunder Horse* rode aback the MV *Blue Marlin* around the Cape of Good Hope, travelling 30,000 kilometres and arriving in the Gulf two months later.

In July 2005, six years after BP discovered the field, production was almost ready to be started up. But the Gulf is not only famous for holding some of the world's richest oil reserves; it is also prone to a yearly battering by hurricanes. Hurricane Dennis was the first major hurricane of the 2005 Atlantic season, the most active one on record. On hearing news of its path towards *Thunder Horse*, BP had evacuated the platform. Picking up in intensity as it made its way towards the US coastline, Hurricane Dennis passed only 230 kilometres from *Thunder Horse* with wind speeds of up to 220 kilometres per hour. Now, as the storm cleared, the hulking mass of steel could be seen listing into the sea.

Back at the National Maritime Museum, Bob finished his speech and sat down. His phone was constantly vibrating, but he could not leave the table to find out what was happening. I had decided not to tell him what I knew until after dinner. A rescue effort could not be launched until the sea had calmed and we could get access to the platform; two or three hours would make no difference. As we were walking out of the door I told him of my conversation with Tony Hayward: 'five billion dollars of investment could be sinking into Davy Jones' Locker'. 'I thought something was up,' he said. 'I'd better go and make a phone call.'

At first we could not understand what had happened: *Thunder Horse* had been designed to survive a 'once-in-a-hundred-years' storm.[25] Sixty thousand tonnes of steel had been brought to its knees, but it was not the weather that was to blame. *Thunder Horse* was already listing at 16 degrees before the hurricane hit and the storm waves only served to worsen the situation. Both mechanical failure and human error had led to malfunctions of the hydraulic control system that keeps the great weight of the platform in balance by moving water between ballast tanks. After several tense days of enquiry and action, *Thunder Horse* was resurrected and since then has stood unmoved by a yearly battering of hurricanes as an example of the immense strength of steel and the scale on which we use and trust it.

Father of the steel industry

When filling up a car at a petrol station, or flicking the switch on a wall socket, most people rarely see how dependent our energy infrastructure is on steel. All the steps in the energy chain, from exploration to production to refining and generating electricity, rely on technology built with iron. But the strength of structures and pipes does not rest on iron alone. If made from pure iron, the atoms would easily slide over each other; *Thunder Horse* would collapse under its own weight. The strength of steel lies in getting the right balance between iron and carbon. Pure iron is soft, but adding carbon breaks up the lattice of iron atoms so that the atoms can no longer easily slide past each other, thus producing hard steel.[26] Add too much carbon, however, and the iron, called cast iron, becomes brittle and shatters when it is struck.

For centuries, steel was only produced in batches by expensive processes which lacked scale. However, in 1856 the chance discovery of Henry Bessemer, an English inventor, led to a process which carefully controls the balance of carbon and iron on an industrial scale. This invention, which is still used today, has had the single greatest impact on the development of the modern steel industry. As with many developments in the iron and steel industry, the Bessemer process was born out of a need for better armaments.[27] In 1854 Bessemer met with Napoleon III, who wanted a superior quality metal so as to improve his artillery pieces. For Bessemer, this 'was the spark which kindled one of the greatest revolutions ... I made up my mind to try what I could to improve the quality of iron in the manufacture of guns.'[28]

Bessemer's breakthrough came in the summer of 1856. One day he opened the door of his experimental blast furnace and noticed some pieces of 'pig iron', a type of iron of high carbon content, sitting on one side without having melted. The temperature must not have been high enough, he thought, and so he let more hot air into the furnace. Half an hour later, to his surprise, the pieces appeared unchanged. He grabbed an iron bar to push them into the bath of molten metal, but discovered that they were not pig iron, but thin shells of pure iron, the carbon having been almost entirely removed. By chance, air had blown through the molten pig iron, raising its temperature and removing impurities. An outside heat source had always been thought necessary to keep the temperature of the furnace high enough to stop the molten iron solidifying. What if, Bessemer wondered, by simply blowing cold air through the molten metal, he could convert an entire crucible of pig iron into pure iron?

So he built another experimental convertor with six pipes at the bottom of the chamber through which to pump air. He opened the valves and air began to push up through the molten pig iron. Bessemer describes what happened next: 'All went on quietly for about 10 minutes ... But soon after a rapid change took place; in fact, the silicon had been quietly consumed, and the oxygen, next uniting with the carbon, sent up an ever increasing stream of sparks ... followed [by] a succession of mild explosions, throwing molten slags and splashes of metal high up into the air, the apparatus becoming a veritable volcano in a state of active eruption. No one would approach the convertor to turn off the blast ...'[29]

After the eruption had subsided, Bessemer poured the molten metal into a pan. It cooled and set into a solid bar. He took a carpenter's axe and struck the bar with three sharp blows. Each time it sank deep into the soft metal, gouging but not shattering or splintering, as would be expected from brittle cast iron. The violent reaction had kept the temperature high in the convertor, without the need for any external heat supply. He was left with a pure low-carbon form of iron: Bessemer Steel.[30] Bessemer's innovative process is now the basis of all steel making. It produced a material that was not only stronger, tougher and more malleable than wrought iron, but could be made in a fraction of the time and, more crucially, at a fraction of the cost. The traditional steel-making process, involving the slow heating of bar iron alongside charcoal, took ten days and cost over £50 a tonne (around US $6,000 in today's money[31]). His process made a tonne at a tenth of that cost. Before that process, steel was so expensive that it could only be used for small, treasured objects such as swords, cutlery and valuable tools. Now, ships, bridges and railways, steam boilers and all sorts of machinery could be constructed with cheap, strong and abundant steel. Even the simple nail could now be made quickly and cheaply from steel without the need for lengthy and arduous forging.[32]

The Bessemer process soon spread around the world. Alfred Krupp was among the first to buy a licence and, by 1867, he was operating the largest Bessemer steelworks on the continent with eighteen convertors.[33] Steel production increased most dramatically when the process was rolled out in the US. In 1892, the US had grown its steel output to four million tonnes a year. It would, according to a *Times* article in 1893, take more than three years' production of all the gold mines in the world to pay for one year's production of Bessemer Steel.

Bessemer was not an ironmaster, but an inventor, engineer and businessman in a very broad sense.[34] He believed that his discovery resulted from the fact that he was not steeped in the traditional practices of making iron. The notion that a blast of cold air could purify molten iron without it solidifying was, on first hearing, ridiculed by many. His earliest and most profitable success resulted not from iron but from a request from his elder sister to help her decorate a book of flower paintings of the many tulips and chrysanthemums cultivated by their father. To do this, he went to a shop in Clerkenwell, London, to buy some 'Gold Powder', made from bronze. Returning the next day to pick up his purchase, he was surprised at its high price of seven shillings for an ounce (about US $40 today). Bessemer was certain he could invent a cheaper way to produce it, and so he did just that without any prior experience in the field. His success in making the powder gave him the confidence and finances with which to pursue a career as an inventor and engineer. As Bessemer would later recount, the request from his sister was 'fraught with the most momentous consequences to me; in fact, it changed the whole current of my life, and rendered possible that still greater change which the iron and steel industry of the world has undergone'.[35]

Sugar-cane juicers, solar furnaces and diamond-polishing machines were all among Bessemer's inventive interests. An inspired genius tamed by a shrewd business mind, he based his inventions on the simple principle of delivering a product that consumers wanted but at a lower price and higher quality than anything that already existed.

Bessemer's success in the iron industry led him to become a founding member and subsequent President of the Iron and Steel Institute, predecessor to the Institute of Materials, Minerals and Mining in London.[36] The Institute was formed in 1869 as a 'closed shop' for British Victorian ironmasters who were worried about competition from Europe. On the walls of the Bessemer Room in the Institute hang numerous awards given to him in recognition of his invention. Bessemer became a fellow of the Royal Society and was knighted. The Bessemer Gold Medal, introduced during his presidency, is still awarded for outstanding contributions to the iron and steel industry. But the medal is one of the few reminders of Bessemer's name that remains today.

Although Bessemer was lauded in his time, he is now largely forgotten; he lies buried beside his wife in a quiet and unvisited corner of West Norwood Cemetery near their former south London home. Despite being one of the world's greatest inventors and engineers, Bessemer is not thought of alongside Thomas Edison, James Watt or the Wright brothers. Perhaps it is because he is the inventor of a process, rather than a product.[37] We remember world-changing objects, such as light bulbs, steam engines and aeroplanes, and also, by association, their inventors. Similarly, we remember the people, such as the amoral arms manufacturer Alfred Krupp, who have changed the world using Bessemer's process. And we also remember one man who did more than anyone else, in his time, with and for steel. Andrew Carnegie was another President of the Iron and Steel Institute and one of America's most famous industrialists.[38] In the Bessemer Room at the Institute, his portrait hangs opposite that of Henry Bessemer. Under the roof of this aged institution, the inventor and the manufacturer stand as equals, but outside we remember only Carnegie who, using the Bessemer process, became the richest man in the world.

Andrew Carnegie's steel empire

It was oil, rather than iron, that created Carnegie's first fortune. In 1859, Edwin L. Drake made his momentous discovery of oil in Titusville, not far from where Carnegie was working on the Pittsburgh Railway lines as a telegraph operator and assistant to Superintendent Thomas A. Scott. The discovery brought a surge of wildcat prospectors to the area. Among the more respectable businessmen in this first wave was William Coleman, who had already made a small fortune in iron manufacturing and coal mining. In 1861, by which time Carnegie had become Superintendent of the Pittsburgh Railway, Coleman invited him to invest in his oil concern. Carnegie received a return that was many times his initial investment and which, in true entrepreneurial spirit, he then invested in his own business.

Carnegie decided to take advantage of his relationship with Scott and Edgar Thomson, the new President of the Pennsylvanian Railway. Many of the wooden railroad bridges had fallen into disrepair during the Civil War and were now rotting. They needed replacing with iron bridges, and Carnegie was just the man to do this. Along with Thomson and Scott, who were still employed by the railroads, Carnegie formed a new company to build iron bridges. Corruption and cronyism were widespread and accepted ways of doing business. It was an environment in which Carnegie flourished; he secured lucrative contracts for his company through friends and business associates, starting with Thomson and Scott.

Henry Bessemer and Andrew Carnegie first met in 1872 when, while on holiday in Europe, Carnegie had visited Bessemer's new steel works in Sheffield. Unlike Bessemer, Carnegie was not an engineer; his skill was in putting proven inventions to their best use.[39] Bessemer became Carnegie's technological brain and Carnegie became Bessemer's salesman. The success of the steel works, and the high-quality steel it produced, was proof to Carnegie of the potential of the Bessemer process. He decided to invest with a recent fortune made by speculating in bonds.

In Pennsylvania there was a clear market for steel. During the Civil War, Thomson had become dismayed by the poor quality of rails on which the Union forces relied for transportation. Made from cast iron, the brittle lines would frequently break. Steel would have been better but it was just too expensive at the time.

Carnegie returned to Pittsburgh and embarked on the construction of a new state-of-the-art steel works, with his old friend William Coleman, to supply the railways. Though there were eight Bessemer steel manufacturers in the US in 1872, none were in Pittsburgh. Two years later, the Edgar Thomson Steel Works opened; from that point on the growth of Carnegie's steel empire became unstoppable.

When demand slowed, Carnegie would increase, rather than reduce, the output of his steel mills. He would take a contract whatever the profit margin, beating the competition down using his economies of scale. His steel mills were always the biggest, the most automated and hence the lowest cost. He would immediately reinvest profits to expand and modernise his steel empire. He also integrated his business horizontally and vertically, buying up rival plants and bringing coke works and iron ore mines under the umbrella of his steel company. Carnegie's success came from business skills that we would recognise today. But it also came at the expense of his workers. In the drive for profits he reduced wages and increased working hours. These actions culminated in a strike at the Homestead Steel Works.

In July 1882, the labour contract between Carnegie and his workers at the Homestead Steel Works was due to expire. The skilled workers at Homestead were being paid more than the industry average and Carnegie saw an opportunity to save on labour costs. He also wanted to reduce the influence of the Amalgamated Association union at the mill. At the time the union, whose membership consisted of skilled metalworkers, held a powerful bargaining position. Carnegie saw this as restricting progress and profits and so Henry Clay Frick, whom Carnegie had placed in charge of operations at the plant, cut the workers' pay. The union would not agree; the steel industry was doing well, and they wanted a share. When the existing contract expired with no new agreement being reached, Frick locked the workers out and built barbed-wire fences and sniper towers to fend off any backlash. He also enlisted a security force, but, as it made its way towards the plant, it was met by the striking workers. Tensions mounted and shots were fired. Many union members were killed. Eventually order was restored and Carnegie got lower wages and removed the union from the plant, but at the price of a tarnished reputation.[40]

As Carnegie's business continued to grow, mile after mile of steel railway poured out from his mills. The railroads were the greatest consumers of iron and steel as more goods and more people began moving from the east to the west of the US. The US steel industry, which for years had lagged behind Europe's, expanded explosively. The amount of steel rolled in 1890 was more than three times greater than ten years earlier. By 1900 output had almost trebled again to more than eleven million tonnes, at which point Carnegie's mills alone produced more steel than the entirety of Great Britain. Under Carnegie's leadership, the iron age gave way to the steel age.

Iron conveys wealth and power not only to nations, but also to the individuals who pioneer its production. Carnegie, the child of a poor hand-loom linen weaver from Dunfermline, Scotland, arrived in the US in 1848 practically penniless. By 1863, at the start of his entrepreneurial career, Carnegie's income was around US $50,000 (almost US $1 million today). In 1901, Carnegie sold his steel empire for US $480 million (US $13 billion today) to J. P. Morgan in what was the largest commercial transaction of its day to consolidate into US Steel.[41] Carnegie became the richest man in the world.

Andrew Carnegie's 'Gospel of Wealth'

In 1986 I was invited to sit on the board of the Carnegie Mellon University Business School by Elizabeth Bailey, the Dean of the School. Elizabeth, an economist, was also an influential member of the Board of the Standard Oil Company of Ohio, where I was the Chief Financial Officer.[42]

I remember pausing and thinking as I scanned the letterhead. *Carnegie Mellon University*. I knew a little about Andrew Mellon: a nineteenth-century American banker who later became the Secretary of the Treasury. I knew a lot more about Andrew Carnegie; his name was everywhere: Carnegie Hall in New York; the Carnegie Libraries across America and Great Britain; and the Carnegie Foundation for Peace. The Carnegie Mellon Business School was one among many prestigious institutions to which Carnegie, through his philanthropy, has attached his name. We may have forgotten Henry Bessemer, whose steel-making invention underpinned

Carnegie's success, but by giving away his fortune Carnegie ensured that we still remember him today. It was all part of his grand vision, which he described in his 1889 essay, 'The Gospel of Wealth'.[43] That immediately caused an outcry among well-heeled capitalists. Carnegie planned to give away his fortune and urged his fellow tycoons, such as J. D. Rockefeller, Jay Gould and J. P. Morgan, to do the same.

At that time, the wealth generated from the industrial revolution in the US was exacerbating the divide between the rich and the poor. Rockefeller became the world's first nominal billionaire in 1916, when many Americans still lived in slums with no running water or electricity. Carnegie believed that this disparity in wealth demanded action 'so that the ties of brotherhood may still bind together the rich and poor in harmonious relationship'.[44] Carnegie's call for wealth redistribution has resurfaced in recent years. Today the top one-tenth of a per cent of all earners in the US take home about 8 per cent of the country's total income.[45] Contemporary billionaires, including Warren Buffett and Bill Gates, ask why the super-rich are not taxed more, and, like Carnegie, support giving money back to society through targeted philanthropy. Just as in the days of the strike at the Homestead Steel Works, sections of the public feel intense anger towards the rich, towards business and towards the political class.

The duty of the man of wealth, Carnegie claimed, was to set an example by living a modest and unostentatious life. All surplus revenues, he wrote, should be considered 'simply as trust funds, which he is called upon to administer … to produce the most beneficial results for the community'.[46] The wealth of the individual had been created by the wider community, and so it should be returned to them. This is the great paradox of Carnegie. His desire to give back all that he earned to society 'encouraged him to be an even more ruthless business man and capitalist … the more he earned, the more he could give away'.[47] Cutting wages and increasing workers' hours were all part of his broader philosophy; he was convinced that he had a moral obligation to increase profits, rather than improve the welfare of his workers. He believed that his 'superior wisdom and experience' enabled him to administer this wealth 'better than [workers] would do or could do for themselves', but he also believed that he should only help them to help themselves.[48] In this way he set a trend that exists to the present day; his philanthropy was designed for people's self-improvement rather than subsidising them day to day. As a result, he invested heavily in education, building libraries and providing free tuition for Scottish university students.[49] He also opened scientific research institutions, music halls, art galleries and even a natural history museum in Pittsburgh. In 1900, he founded the Carnegie Technical Schools, which would later merge with the Mellon Institute of Industrial Research to become Carnegie Mellon University.

His work had a great impact on Rockefeller, who wrote to him after the Carnegie Library opened in Pittsburgh: 'I would that more men of wealth were doing as you are doing with your money; but, be assured, your example will bear fruits, and that time will come when men of wealth will more generally be willing to use it for the good of others.'[50] Yet, in private, he discussed the apparent vanity in Carnegie's philanthropic donations. The manager of one of Rockefeller's charitable trusts wrote to him that Carnegie gives money 'for the sake of having his name written in stone all over the country'.[51] In comparison, Rockefeller's donations were discreet, perhaps so that he could not be accused of trying to curry favour in the face of Standard Oil's trouble in the courts.[52]

In 1910 Carnegie founded the Carnegie Endowment for International Peace, which is still active today. For a long time Carnegie had been an outspoken campaigner for world peace, opposing foreign intervention by the US and the UK. On the eve of the First World War he founded the Church Peace Union. By promoting moral unification and leadership, he hoped the Union would put an end to war for ever. Iron, or more accurately the profits from iron, was again being used as a force for peace. By the time of his death in 1919, Carnegie had given away 90 per cent of his wealth. He arranged for the rest to be distributed through the Carnegie Corporation of New York.

Despite the events of the strike at the Homestead Steel Works, we remember Carnegie as a benevolent philanthropist. Even Henry Clay Frick, the antagonist of the tragedy, is now remembered by a glorious collection of art on Fifth Avenue in New York.[53] Men are remembered by their ends, by their ultimate generosity, rather than the process by which they achieved those ends. As we look at the Vermeers, Holbeins and Rembrandts in the Frick Gallery we think of beauty and benevolence and not of the innocent lives lost at Homestead.

The sky's the limit

In mid-1971 I had been living in New York for about six months when a friend of mine, a professional cellist, invited me to a concert at Carnegie Hall, which sits across Central Park from the Frick Collection. We went to listen to Paul Tortelier, the famous French cellist, playing *Variations on a Rococo Theme*, a piece by Tchaikovsky in the warm and colourful acoustics of the hall. When the hall opened in 1891, with a concert conducted by Tchaikovsky, it quickly became a New York City icon. The striking Italian Renaissance-style façade of terracotta and iron-spotted brick was as fascinating to many visitors as the illustrious artists performing inside. The hall was designed by the architect William Burnett Tuthill, himself an amateur cellist. He decided against the use of steel support beams, but as a consequence had to build concrete and masonry walls several feet thick. Stone simply cannot carry the same load as steel. It is the thickness of the walls, along with the smooth elliptical hall interior, that lends Carnegie Hall its fine acoustic properties (and helps to drown out the rumble of traffic from 57th Street and Seventh Avenue outside). In 1891 it was already being overshadowed by a new breed of building, shooting up across the city: the skyscraper.[54] Just six-storeys high, Carnegie Hall was dwarfed by headquarters of the *New York Times* and *New York Tribune*. Ironically, it was the availability of cheap and abundant steel produced by Carnegie that supported this growth.

As early as in ancient Greece, iron had been incorporated into buildings so as to improve stability, but never had an entire load-bearing structure been fashioned entirely out of iron. In the 1820s cast-iron columns and beams were introduced into buildings in Chicago and New York. Architects and engineers were impressed by its compressive strength and durability and soon whole buildings' fronts were being made from cast iron. As an added bonus, unlike wood, iron did not burn but it did, however, melt. In the extreme heat of Chicago's Great Fire of 1871, building fronts buckled and collapsed and iron fell out of favour. Soon stronger and safer steel, pouring out of Carnegie's steel mills, enabled steel skyscrapers to rise upwards.

Architects and the public alike still worried whether they could trust steel to support such gigantic structures, but there was a need to find new space in the increasingly cramped cities. Great social and economic changes were sweeping New York. Immigrants were arriving daily from Europe, sailing past the steel-framed Statue of Liberty. Trusts and corporations were growing quickly and all wanted office space in the financial and business capital of the US. Land prices were rising rapidly and the only direction in which to expand was upwards. Traditional stone was just not suitable. The taller the building, the greater the downward weight and therefore the thicker the base had to be. Early plans for the nine-storey masonry structure of the now demolished Tribune Building show basement walls that are two metres thick, a huge waste of valuable floor space. For architects to reach above six floors they had to rely on steel. Between 1870 and 1913, New York was raised from a city of six-storey buildings to one of fifty-storey skyscrapers.

When I arrived in New York on St Patrick's Day, 17 March 1971, I hated it. I did not know anyone, the hotels were shabby and the people were rude. But, becoming a resident of Greenwich Village, I soon grew to love the city. Fast-paced and full of interesting people, New York was, and still is, one of the most exciting places in the world. Like other great world cities such as Venice, Tokyo and London, New York is a unique and unmistakable urban environment. In particular, I remember being struck by the city's fantastic architecture. Elegant brownstone town houses are interspersed with imposing gothic and art deco skyscrapers. The city would rise up around you on all sides to a vertiginous height. It was very different from anything I had seen in Europe.

At weekends in New York I would cycle around the city. Leaving my apartment off Washington Square, I headed south through the then almost deserted district of Soho, which is the site of the most significant and largest collection of nineteenth-century cast-iron buildings. I would head back north and just above Houston, in a rather grimy and run-down office area, one building stood out. It was called the Flatiron Building and it symbolised to me the growth of New York, over the last one hundred years, into the thriving metropolis of the 1970s. As its name suggests, the shape of the building resembles an old clothes iron, with a stark triangular cross-section that rises up to a height of 87 metres. The twenty-two-tiered structure may seem squat in comparison to New York's modern-day mega-structures, but, on completion in 1902, the Flatiron was a significant feat of engineering.

It was built by the George A. Fuller Company. George Fuller was an architect and pioneered the building of skyscrapers in the US. In 1900, long before the first steel struts of the Flatiron were raised, Fuller died and the presidency of the company was assumed by his son-in-law, Harry Black. For some time he had had his eyes on the small triangular plot of land at the intersection of Broadway and Fifth Avenue. The site would be perfect for the company's new headquarters, acting as an advertisement for his rapidly growing business. Six months after Fuller's death, with the company now in his control, Black bought the land. The prime motivation for building upwards was profit. An editorial in *Life* magazine in June 1901 declared, 'here in New York the price of land determines the height of the building'.[55] And land prices were very high. Only above the tenth floor would investors break even on the land purchase. The tiny plot of land, just 9,000 square feet, had cost Harry Black $2 million ($55 million today).

As a result, aesthetics and economics were often at odds as skyscrapers began to dominate the city skyline. Many worried that the city would be overshadowed by unimaginative yet towering monstrosities. The Flatiron was designed to make the most out of the oddly shaped triangular plot of land, but Harry Black would often argue with the architect, Daniel Burnham, over the building's soft, curved edges. Why, Black would demand, were 93 square feet of valuable floor space being wasted? The Flatiron soon began to rise up from the wedge-shaped site. Thousands of steel columns, joists, struts and rivets arrived truck by truck and were put together much like a child's constructor set. Finally, the builders, standing on wooden platforms that hung down the length of the building, laid the terracotta-tile skin. In June 1902, it was complete.

The Flatiron was different from any other skyscrapers in the city. Many architectural critics regarded the building as a feat of engineering but a failure as a work of art. But the building was embraced by the public and artists alike. The *New York Tribune* reported in 1902 that the Flatiron would attract crowds of 'sometimes 100 or more' and was painted and photographed more than any other building in the city.[56] Symbolically modern, the Flatiron appears to change shape depending on where you were standing. On a winter's day in 1902, photographer Alfred Stieglitz stood looking up at the Flatiron and, pressing the shutter button, created a memorable image.[57] 'With the trees of Madison Square covered with fresh snow, the Flat Iron impressed me as never before,' he wrote. 'It appeared to be moving towards me like the bow of a monster ocean steamer, a picture of new America still in the making ... The Flat Iron is to the United States what the Parthenon was to Greece.'[58]

The unique shape and location made the Flatiron another, and instant, New York City icon, but it was far from being the tallest building in New York. For one reason, its height was restricted by the strong winds that blew down 23rd Street, which bordered the southern edge of the Flatiron. Build too high, so the architects worried, and the building might be brought tumbling down. Locals placed bets on how far the debris would fall if this were to happen. It was an unlikely event as the Flatiron was designed to withstand four times the maximum wind force that it would ever come up against. In part, this strength lay in the building's odd shape. A triangle is the strongest of geometric shapes as it is a self-supporting structure: applying pressure at one point creates greater resistance at the other two points. Soon after the first tenant moved in, a 100 kilometres per hour windstorm braced the city, but inside the Flatiron not a vibration was felt. One tenant claimed that not even the filament inside his desk lamp shook. In fact, it was the Flatiron that held control over the wind. The shape of the building would channel the wind into blustery down draughts that could raise the skirts of female passers-by. Men would loiter around the tower in the hope of catching a rare glimpse of ankle, the police calling out '23 skidoo', since the Flatiron is on 23rd Street, to move the chancers along. One local dressmaker even created a 'wind-defying' skirt.[59] Hats blew off heads and umbrellas were turned inside out, but the wind could also create more serious havoc. On the sur-rounding streets shop doors were blown open and plate-glass windows smashed. One February afternoon in 1903 a fourteen-year-old messen-ger boy, who was attempting to make it around the prow of the Flatiron towards Broadway, was blown into the middle of Fifth Avenue and killed by a passing motorcycle.

As more skyscrapers rose around the Flatiron, the sharp wedge shape lost its propensity to channel wind and the gusts died down. Cycling along Broadway in the 1970s, the Flatiron was no longer as imposing, overshad-owed by the nearby Metropolitan Life Tower. But the building contin-ues to attract people to the area and is a constant reminder of a period of unparalleled growth in the city.

Skyscrapers are built to accommodate more people in a city centre; building upwards is just a more efficient use of the land. Iron's strength and abundance underpins this concentration of the human population into small pockets dotted about the world. These hives of activity foster innovation and so support human progress, raising humanity from simple rural communities to prosperous urban societies.

Skyscrapers were also built as symbols of economic and political power. Some of the first inhabitants of New York's pioneering buildings were the executives of the corporate trusts, such as Standard Oil and US Steel (formed from the Carnegie Steel Corporation and several other major pro-ducers), which resulted from the industrial revolution. Prestige was vitally important to Rockefeller, Carnegie and the other American robber barons and so these buildings were beautiful as well as big.

The World Trade Center's Twin Towers were the tallest buildings in the world when they were completed in 1971. For thirty years they stood as symbols of America's economic eminence, and this is why they were chosen as targets in the terrorist attacks on 11 September 2001. The Twin Towers went down, and with them 200,000 tonnes of steel. On the day of the attacks, New York City mayor Rudy Giuliani declared: 'We will rebuild. We're going to come out of this stronger than before, politically stronger, and economically stronger. The sky-line will be made whole again.' The Ground Zero site is being restored to its former standing as a centre for global economic activity; on completion, One World Trade Center will be the tallest skyscraper in America.

The Iron Lion of Cangzhou

At the beginning of 2012 there were, for the first time, more people living in cities than in rural areas. Nowhere is this dramatic shift more apparent than in Asia. The Petronas Towers in Kuala Lumpur, Malaysia, are the biggest twin towers ever built, while Dubai's Burj Khalifa is the world's tallest skyscraper, standing 830 metres high. Iron and steel have made this possible; China is now the world's biggest consumer of iron ore.

Every time I visited China in the 1980s and 1990s, new ramshackle and purely functional apartment and office blocks would have sprung up with extraordinary speed. In this period of rapid economic growth, the pragmatic prevailed over the aesthetic; there was little time to consider how to create a sustainable and beautiful city environment. Today, many of these early buildings have been knocked down and replaced with piercing skyscrapers. China's cities are getting taller and sprawling faster than anywhere in the world. The simple, elegantly curved Shanghai World Financial Centre and the oddly pleasing Beijing headquarters of China Central Television, colloquially known as 'the big-shorts', are Flatirons of the twenty-first century.

These are not the first iron colossi that China has built. In the small city of Cangzhou, 240 kilometres to the south of Beijing, stands the Iron Lion. It is over five metres tall and weighs around 50 tonnes. Originally, the Lion is thought to have lived inside a Buddhist temple, with a bronze statue of Bodhisattva Manjusri riding in the lotus flower on his back. Today it is battered and broken. The Lion's tail went missing sometime around the turn of the seventeenth century, and in 1803 a storm toppled the statue, chipping the beast's snout. It may have seen better days, but the hulking cast-iron mass persists as evidence of the phenomenal Imperial Chinese iron industry that existed over a thousand years ago. When it was cast in AD 953, the production of iron in China was far greater than anywhere else in the world. Iron output increased sixfold between AD 800 and 1078 to 115,000 tonnes, almost as much as the whole of Europe would produce in

1700. China's growing iron industry was just one face of the great social, political and economic changes that were sweeping through the country during the late Tang and Song dynasties. Joseph Needham, the historian of Chinese technology, writes that 'China in 1000 had more in common with China of 1900 than it had with the China of 750'.[60]

With this prodigious output of iron, China cast the tools and weapons that it used to become a dominant global power at the beginning of the second millennium A D.[61] Up until 1700, China had the world's largest and most efficient iron industry, which during the nineteenth and twentieth centuries declined. During this period, a Chinese poet, Ji Ruiqi lamented this change:[62]

> *Thinking of ancient flourishing glory*
> *I sigh for the changes of our times*
> *But the Iron Lion still stands,*
> *While halls and palaces have turned to thorns and brambles.*

In Europe and the US the industrial revolutions were creating increased productivity and growth; scientific advances had led to an array of innovative low-cost iron- and steel-making techniques. China could now simply not compete. In the late 1950s, Mao Zedong was determined to reverse the decline of its iron and steel industry during the Great Leap Forward. He wanted China to produce more steel than Great Britain but the pots and pans he had collected and melted down produced useless pig iron.[63]

Under the rule of the British Raj, the economy in India was also stalled. Jamsetji Tata, a successful cotton and textile industry entrepreneur, wondered what could be done to reverse this position. With his already-made fortune Jamsetji decided to pursue his own vision: the development of India.

The house of Tata

Jamsetji believed that four ingredients were necessary for industry to flourish in India. First, technical education and research were needed to reduce India's reliance on foreign technology. Second, he saw that hydroelectricity would utilise India's huge supply of water to generate cheap electricity. Third, he made plans to build a grand hotel, to attract the wealthy international elite to India. Finally, and most crucially, Jamsetji wanted to produce steel, 'the mother of heavy industry', for the building of railways and cities.[64]

The notion that India could succeed in making steel to British standards was ridiculed. Sir Frederick Upcott, Chief Commissioner for the Indian Railways, offered to 'eat every pound of steel rail they succeed in making'.[65] Behind this arrogance, however, was the fear of competition, a fear which for a long time had caused the British Raj to hold back industrial development in India. It was not until the 1890s that the Raj began to support the development of the iron industry in India, when Great Britain saw itself slipping behind the booming iron industries in Germany and the US.

Jamsetji finally earned government support when he met Lord Hamilton, the Secretary of State for India, in 1900. Now sixty and wealthy, Jamsetji explained to Hamilton that his efforts to develop a steel works were for the improvement of his home country rather than personal profit. Hamilton told him that he would have full government backing. At once, Jamsetji left for the US, where he sought the advice of Julian Kennedy, America's foremost steel expert and manager at Carnegie's well-known Edgar Thomson Steel Works. Then he hired a geologist to seek out suitable iron ore reserves, a task he had started almost two decades earlier.

In 1904 Jamsetji Tata died. Of his four great plans for India, only the Taj Mahal Hotel had been completed, built with ten pillars of spun iron that still hold up the ballroom ceiling today. The execution of his vision was continued with equal determination by his sons, Sir Dorabji and Sir Ratanji Tata. Three years later, Dorabji found the perfect site for the Tata steel plant at the small village of Sakchi, which had had a good supply of iron ore, water and sand. Sakchi was later renamed Jamshedpur in honour of Jamsetji. In 1912, the Tatas' plant produced its first ingot of steel, marking the birth of India's new iron industry. During the First World War, Tata Steel exported 2,500 kilometres of steel rails, leading Dorabji to comment that if Sir Frederick Upcott had fulfilled his promise to eat every quality steel rail produced he would have suffered 'some slight indigestion'.[66]

The Tatas' first steel mill might never have been built if it had not been for their nationalist vision. Seeking investment for the Sakchi steel mill, the Tatas appealed to their fellow Indians to invest in a project that would help industrialise India. One observer recounts how '... from early morning till late at night, the Tata offices in Bombay were besieged by an eager crowd of native investors. Old and young, rich and poor, men and women they came, offering their mites; and at the end of three weeks, the entire sum required for construction requirements ... was secured, every penny contributed by some 8,000 native Indians.'[67]

At all Tata's industrial enterprises, the wellbeing of the workers was a priority. Jamsetji introduced humidifiers, fire sprinklers, sanitation disposal and water filtration into his early cotton and textile mills. He was also a pioneer of pension funds, accident compensation and equal rights. Unlike Carnegie, the Tatas did not pursue profits at all other costs; they believed that a business that supports the development of a nation must also support the health and wellbeing of its people. A century before Western business had defined corporate social responsibility, enlightened self-interest and the triple bottom line, he had already got there.[68] Tata knew that consideration for society must be integral to every aspect of operations. When shareholders complained of the great expense going into building workers' accommodation on the site of the Tatas' first steel mill, Sir Dorabji would tell them: 'We are not putting up a row of workmen's huts in Jamshedpur, we are building a city.'[69] In 1895 Sir Dorabji wrote: 'We do not claim to be more unselfish, more generous or more philanthropic than other people. But we think we started on sound and straightforward business principles, considering the interests of the shareholders our own, and the health and welfare of the employees, the sure foundation of our success.'[70]

There are few more eloquent expressions of the proper relationship between business and society. Even today many businesses have failed to catch up with Tata. Some still follow American economist Milton Friedman's mantra that 'the only business of business is business', and believe they can ignore the external world.[71] In doing so, they risk their reputation, their licence to operate and their customers. Today, as the scrutiny of business becomes more intense than ever, and as governments call on business to do more and more beyond their core activity, we would do well to refer to Tata's example, and to observe the success it became, not only for his dynasty but for India as a whole.

In *The Argumentative Indian*, Nobel prize-winning economist Amartya Sen considers the role that our sense of identity and social motivation play in the determination of our economic behaviour. 'Within the limits of feasibility and reasonable returns,' he writes, 'there are substantial choices to be made, and in these choices one's visions and identities could matter.'[72] For the Tatas, a nationalist vision of India's industrial development led them to behave quite differently in their business practices from the American robber barons.

Like Carnegie, Jamsetji Tata also gave money to support the needs of his nation. With the large fortune that Ratanji and Dorabji inherited, they set up philanthropic organisations, with a focus on education, health and founding institutions of national importance. They continue to hold a large shareholding in Tata today. Jamsetji Tata is remembered in India in the same way as Carnegie is remembered in the US. They differ in how they acquired their wealth. Carnegie sought to amass as much wealth as possible, whatever the cost to his workers, so that he would have more to give back to society, while, for the Tatas, their personal wealth and that of India were one and the same. Both used iron to develop the industrial power of a nation, to form the foundations of a long-lasting business and, ultimately, to give back to society.

Their approaches were successful in their own time and place, but neither would be entirely successful today. In the lawless early days of the American industrial revolution, cronyism and corruption were the only ways for a business to survive and grow. Today such behaviour would provoke an immediate and powerful public backlash; a business would not grow or survive. Even the Tatas' pioneering and enlightened approach to business is not wholly sustainable in today's business environment. Dorabji's decision to build 'a city', rather than 'a row of workmen's huts', would run contrary to the fundamental business principle of maximising shareholder value.[73] Dorabji's shareholders were the people of India, whereas today shareholders are largely private individuals. In the current business environment, social investment remains a vital aspect of ensuring business sustainability, but there is a need for something more. In the face of growing inequality, philanthropy remains a vital mechanism for redistributing the wealth that individuals acquire through their use of the elements. In Carnegie and the Tatas we find lessons for today.

CARBON

Every year, on the third Saturday in July, I watch the fireworks of Venice's Festa del Redentore, the Festival of the Redeemer, illuminate the sky above the church of the same name. With friends I take a gondola into St Mark's Basin and meander through the festively decorated boats, packed solid save for a narrow passage kept open for gondolas. Even today, the gondola still commands precedence and respect from all other vehicles in the city. We dock close to the old customs house, the Dogana, and admire the magnificent spectacle of lights as they reflect in the still lagoon waters around us.

The Festa del Redentore celebrates the end of the plague that swept through Venice between 1576 and 1577, taking 51,000 Venetian lives, including that of Titian, the great Renaissance painter. Later that year, as the plague subsided, the foundation stone of Il Redentore, the Church of the Most Holy Redeemer, was laid on the island of Giudecca as an act of thanks. A floating bridge was created out of barges between Giudecca and the main island to allow the Doge, the leader of the Republic of Venice, to walk in procession to the church. That tradition is continued to this day, without, of course, the Doge.

Venice grew out of virtually nothing, from a marshy, malarial wasteland to become, by the thirteen century, the prosperous and most recognised commercial centre of the Western world. Few other cities have made more of a contribution to painting, sculpture, architecture and music. The Festa del Redentore is a celebration of the city's enduring spirit during times of adversity. We watch the rockets shoot skywards and explode in jubilant

streams of colour. Atoms of each element flare their characteristic colour: copper blue, strontium red and sodium yellow.[1] But the element of essential importance to the light display is carbon. Without it the rockets would never leave the ground.

Carbon is the fuel supply in gunpowder. In ancient China, where the first fireworks were invented, honey was the source of carbon; the drier the honey, the greater the carbon content. The explosive mixture was reported to burn hands and faces and even burn down houses. Later, charcoal, made from the carbonisation of firewood, came to be the traditional fuel element of gunpowder.[2] Now, as in the first fireworks of the Festa del Redentore centuries ago, carbon is the fuel element that ignites these yearly festivities.

The fabric of which most of Venice is made is threaded through with carbon. The most widely used stone in the city is a form of limestone, calcium carbonate, from Istria, a peninsula which was part of the territories of the Republic. From Istrian stone the great columns, archways and façades of Venice's historic palazzos were crafted. Both easy to carve and resistant to weathering, this carbonaceous mineral dominates the city. Carbon, too, is intrinsic to the paper and ink on which Venice laid down its monumental artistic and intellectual achievements and is embedded in the wooden ships with which Venice waged war and expanded her mighty Republic.[3]

The unusual ability of carbon atoms to bind to themselves to form long chains and rings makes carbon one of the most versatile elements.[4] Carbon creates the structures off which other elements hang to create the great complexities of the world. Carbon is not only the foundation of the Venetian Republic, but also of all human civilisations and humans themselves. Carbon is the backbone of DNA, the genetic code from which our bodies are built, repaired and replenished. It is the element of life.[5]

But carbon, in isolation, is mostly useless. Gunpowder would not explode if the fuel consisted purely of carbon, which ignites only at extremely high temperatures. An oxidiser is needed to speed up the burning of the carbon to make the firework go off with a bang. It is the complex mix of carbon and oxygen that fuels the fireworks at the Festa del Redentore each year.[6] Context is almost everything for the carbon atom.

There are very few uses for pure carbon. It appears in graphite, from which pencils are made, and in diamonds. It is used for jewellery and drill bits. Graphene, a pure carbon lattice which is just one atom thick, could soon revolutionise many high-technology industries.[7] But overwhelmingly carbon needs other elements to make a difference to the world.

An outbreak of the plague was one of the first nails in the coffin of the Venetian Republic as it began to decline in the sixteenth century. In May 1797, the Republic came to an end when Napoleon Bonaparte's troops surrounded the lagoon entrance. The Great Council of Venice had little choice but to surrender. With Zen-like calm, the Doge removed his ducal 'corno' and passed his white linen cap to his valet, saying simply: 'Take it, I shall not be needing it again.'[8]

After the Republic fell, the city reinvented itself by turning to a trade of a very different kind: becoming an attraction for more and more visitors. Now, throughout the summer the historic walkways and squares are packed with tourists of all nationalities who come to marvel at Venice's cultural legacy. Despite the modernisation of the Venetian economy, the city itself appears much the same as it did when the Republic fell in 1797. The absence of four-wheeled transport gives the impression that modernisation has passed the city by. However, without the constant stream of aeroplanes landing at Marco Polo airport, or the cars and buses crossing the Ponte della Libertà, the economy would collapse. Much like in the fireworks exploding above the Church of the Most Holy Redeemer, carbon now fuels the economy in the form of hydrocarbons used for transport.[9]

Of all carbon's combinations, hydrocarbons are the most world-changing. Hydrocarbons are a variety of molecules which are made from the elements hydrogen and carbon. Generally, when they are mixed with oxygen from the air and a spark is added, heat is released. This exothermic reaction, with enough oxygen, also releases carbon dioxide and water vapour.[10] In the form of fossil fuels – coal, oil and gas – carbon has granted us simple and abundant energy sources with which to shape the world. The rapid improvement in living conditions and growth in global population seen over the centuries since the Industrial Revolution could not have happened without these fuels. They are the focus of carbon's story.

I begin with coal, the use of which exploded during the Industrial Revolution and placed societies on the road to global economic growth. Then comes oil, first pumped from wells at the end of the nineteenth century and used as a cheap, bright and clean source of illumination that extended the time available for both work and leisure. However, it was in combination with the automobile that oil really changed the world. The automobile gives people the freedom to go where they want, when they want. And beyond that, oil powers the global network of trucks, ships and aeroplanes that transport cargo and passengers and ensure the functioning of our modern society. And finally natural gas, the most volatile of fossil fuels, which for millennia humanity has simply regarded as waste as it was too difficult to use. Now it has found a powerful purpose in electricity generation and is capable of being moved from where it is produced to where it is needed. The end of my story of carbon tackles an important question: are coal, oil and gas, though of great benefit, the stuff which through our enormous consumption will create catastrophic consequences for humankind?

I first began to understand the dramatic ability of hydrocarbons to shape the world in the 1950s as a child living near the Masjid-i-Suleiman oilfields in Iran. My father worked for the Anglo-Iranian Oil Company (later BP), and he would often take me out to visit the oilfields with their 'flares' burning what was then useless natural gas. The intense heat, noise and sulphurous fumes were overpowering. At the age of ten, I remember being taken to the site of Naft Safid Rig 20, a well which spectacularly caught on fire years earlier. The well exploded, spewing hundreds of metres of pipe into the air. The pipe lay where it landed all around the well head. I also vividly remember watching the orange glow of the blowout of the Ahvaz No. 6 oil well burning fifty miles from our house. I was enthralled by the sheer power of oil.

My interest in the oil industry would remain with me for the rest of my life, leading me to join BP in 1966. At BP I tackled the technological and political challenges of oil extraction. I navigated through unstable political terrain of Russia and Colombia, facing oligarchs, drug barons and guerrilla fighters. I saw how, in the struggle for access and control of this highly valued resource, oil brings out the greed and malign spirit of powerful leaders. It occasionally brings out the good. It certainly changed nations who own it and, indeed, it changed the world.

But carbon's story begins millennia before we began to use oil on an industrial scale. In 1969, I was working in Alaska and there I witnessed a natural phenomenon that led me to reflect on humanity's earliest use of hydrocarbons. I was a member of a field survey team, taking measurements in the Brooks Range Mountains across which the Trans-Alaska Pipeline runs. During a storm, I watched lightning strike a bituminous outcrop, setting the exposed veins briefly on fire. It is just possible that, thousands of years earlier, early humans saw the same phenomenon and so learnt about the combustible properties of coal.

COAL

Today, by far the largest consumer of coal in the world is China. There is a striking continuity here, for the earliest known use of coal dates from around 4000 BC in the Shen-yang area of north-east China. Pieces of soft, black rock, known as jet or lignite, scattered about the ground were found by early humans and carved into decorative objects, such as beads and ornaments with which to pierce ears. Jet, being easy to carve and polish, was perfect for creating ornaments. It is still used in jewellery today because of its distinct, high-quality finish. The term 'jet black' originally comes from this rock.

Coal is the heaviest form of hydrocarbon. It has a smaller proportion of hydrogen to carbon than oil or gas. Therefore it is more difficult to burn and, when burnt, it produces a greater proportion of carbon dioxide than these hydrocarbons.[11] Coal comes in different forms; the more hydrogen it contains, the easier it is to ignite. The early inhabitants of China probably saw coal ignited by lightning, or perhaps by the sparks from a fireplace, and then realised that they could use it as a source of heat and light.

By the time of the Han Dynasty, beginning around 200 BC, coal was being used on a large scale for both domestic and industrial purposes. It is not surprising that humanity's first use of coal was in China. Small deposits are found in every province, while the world's largest coal deposit is in the north, centred on the province of Shansi.

When timber, the traditional and most easily accessible fuel, became scarce during the eleventh century, the use of coal expanded dramatically, mostly in the growing iron industries in China. At this very point in history, China became the centre of innovation for the world. It invented gunpowder, the compass, paper and the printing press, the so-called Four Great Inventions.[12] These took many centuries to reach the West.

Putting to use the energy inherent in hydrocarbons, China established itself as the world's leading power. At that time, China seemed to be on the verge of an industrial revolution, but soon after its boom in innovation its coal consumption slowed and its development relative to the West faltered. This led to the 'Great Divergence', the term given to the increasing split between the West's and East's power and wealth through the later centuries of the last millennium.

Why did this happen? Was it simply due to differences in geography and geology? Or were the differences cultural, China emphasising the group over the individual? It is one question Joseph Needham set out to answer in his great work *Science and Civilisation in China.*[13] I like to think that it was at least in part about access to energy, and that contains a lesson for today. Whatever the answer, Needham writes: 'By 1900 China was out of the race, and Western industry dominated the world.'[14]

Meanwhile, Great Britain, one of China's chief competitors, was using its own source of abundant coal to fuel the world's first Industrial Revolution.

Escaping the 'Malthusian Trap'

Writing in 1798, Thomas Malthus observed how, throughout history and across different cultures, living standards had not grown.[15] Humanity existed perpetually on the threshold of basic subsistence. Agrarian societies did develop new technologies to increase food production, but this only led to an increase in the population and reversion to a level of basic subsistence. Economic resources could never outpace population growth; humanity was stuck in a 'Malthusian Trap'. Yet as Malthus was writing, great changes were beginning to shape Great Britain around him. These changes would soon disprove his theory.

Between 1750 and 1850, Great Britain's industrial output grew seven times, while the population grew less than threefold. Living standards rose and rose. There is no simple answer as to why this happened in Great Britain first. But just as the decline of China relative to the West was linked to its stagnating coal industry,[16] the industrial rise of Great Britain was, at its core, dependent on putting to use its abundant coal reserves. Between 1700 and 1830 the production of coal increased more than tenfold. Steam engines replaced manual labour because coal in an amount equal to the weight of a man can produce the equivalent energy of the same man working for one hundred days.

Far more than fuel to create steam, coal is also vital in the production of iron and steel. It is used to smelt ore in furnaces and added to molten ore in the form of coke to remove impurities. Built with bars of iron and powered by coal fires, factories and mills sprung up across the country at an astonishing rate. The small northern town of Manchester in England quickly grew to become a symbol of Great Britain's Industrial Revolution, stretching for miles across the English countryside. Coal smoke would greet visitors from afar, rising from the towering smokestacks of the city's cotton mills.[17] When visiting in 1835, Alexis de Tocqueville, the French historian and political thinker, summed up the ambivalence of the situation: 'From this foul drain the greatest stream of human industry flows out to fertilize the whole world. From this filthy sewer, pure gold flows.'[18] In the south, London emerged as the 'centre of the commercial and political world' and the economy grew at an unprecedented rate.[19]

Great Britain was transformed from a nation of farmers to a nation of manufacturers and factory workers. People swarmed from the countryside into towns to work in the coal-powered mills and factories. For many who arrived in the rapidly growing cities, living conditions were poor. Whole families were crammed into single squalid rooms and death and disease were far higher than in the countryside. This prompted Friedrich Engels, the German philosopher, social scientist and industrialist, to observe that workers suffered more than they gained from the Industrial Revolution.[20] The urban growth continued and by 1851 there were more people living in cities than outside them, a tipping point only reached by China in 2012.

Coal output in Great Britain grew to many times more than the rest of the world, but rapidly growing manufacturing, alongside a rapidly growing population, put great pressure on the mining industry. Already by the end of the eighteenth century, most of the low-lying, easy to access coal seams had already been exploited. To increase supply, miners would have to dig deeper.

In the rush to meet and profit from demand, thousands of lives were lost.[21] Deadly gases hidden in the coal seams suffocated miners or led to explosions in the mine shafts.[22] Children, some as young as eight, worked as 'trappers', sitting for several hours at a time in the cold, dark and damp mining tunnels, periodically opening a ventilation flap to keep air moving down the shafts. They were further abused by being used to haul coal up shafts too narrow to be passed by adults. Working conditions were appalling.

Over time, regulations began to improve poor working conditions. In 1842, laws were introduced, prohibiting underground work for women and children below the age of ten.[23] More laws and enforcement bodies improved mining safety. Developments in technology were equally important; the coal-powered steam engine enabled water to be drained from flooded mines and coal to be mechanically hauled to the surface, while the invention of the Davy lamp in 1815 prevented the ignition of explosive gases in the mine shafts.[24]

It took far longer to clear the thick black smoke pouring from the chimneys of coal-powered factories. Coal is the most polluting of fossil fuels, burning inefficiently and producing high quantities of soot and harmful pollutants, not least carbon dioxide. During my childhood in Cambridge during the late 1950s, coal made its presence felt. When the wind blew in the right direction, the foul smell of sulphur and ammonia would blow through the open windows of our house on Chesterton Road from the nearby town gas works.[25] In the 1950s, large reserves of cleaner natural gas had not yet been discovered in the North Sea, and households still depended on town gas, manufactured from coal, for cooking and heating.

Air pollution in today's industrialising countries is still a major environmental and health problem. Nowhere can this be better seen than in China, which is now the world's largest consumer of coal. Its coal use, once stagnant, is now growing at an unprecedented rate, fuelling the sudden growth of its economy seen over the last thirty years. The Great Divergence is rapidly closing, but as China catches up with the West it is finding that history repeats itself.

China's dilemma

I first visited Beijing, on BP's behalf, in 1979 shortly after Deng Xiaoping opened the door to international commerce. After the death of Mao Zedong and the arrest of the infamous Gang of Four in the autumn of 1976, Deng came to power and began a series of major economic reforms in China, leading it towards a market economy.[26] In the seven years following these reforms, the number of people living in poverty halved from 250 million to 125 million. Each time I have visited the country since, the living conditions of those in the city have seemed to improve. The food gets better and the people get fatter. The change to Beijing's skyline is even more dramatic. Skyscrapers are built three at a time in the same design;

the pace of growth in the city is prodigious, pragmatic and relentless. Fuelled by hydrocarbons, China's booming economy has taken the once poverty-stricken nation to the status of a world economic superpower in a matter of decades. In that transformation, coal is the dominant force: it generates around 80 per cent of China's electricity and provides 70 per cent of China's total energy demand. In 2010, China used almost as much coal as all other countries in the world combined. However, as with the Industrial Revolution in Great Britain, China's prodigious coal use comes with consequences for both health and the environment.

I still remember the acrid, yellow air on a visit to Beijing's Summer Palace in October 2003. I had come to China's capital to meet Gary Dirks, then head of BP China. Taking a break from the chaos of the Central Business District, we headed out to the Summer Palace, one of the city's many beautiful imperial gardens, to continue our discussions. Walking along the 17-Arch Bridge that connects the Palace's Kunming Lake to a small central island, we were debating whether to sell our shares in PetroChina, which we eventually did. Halfway across, neither shoreline was visible; the thick smog had reduced our visibility to only a few metres, creating a welcome sense of privacy for our meeting. Other visitors, many wearing face masks, would appear out of nowhere before disappearing just as quickly into the mist on the other side.

In the winter and spring, certain atmospheric conditions trap pollution within the city which builds up to almost unbearable levels.[27] The true extent of the city's toxic emissions, a choking mix of sulphurous and nitrous oxides, soot and other particulate matter, are then revealed. Coal, more than any other substance, is the culprit of this pollution.

As I walked around Kunming Lake, my throat became raw and inflamed because of the air around us. Inhaling this pollution has grave consequences for health: invisible soot particles can accumulate in the respiratory system, increasing the risk of lung diseases. Over 350,000 people in China are estimated to die each year as a consequence.

The health risks are even greater for those mining China's coal. Black lung disease affects hundreds of thousands of Chinese miners. Mine collapses and explosions are common: in 2005, more than two hundred people were killed in just one gas explosion at the Sunjiawan mine in Liaoning.

More than 2,500 coal-mining deaths were reported in China in 2009, and over the previous decade the number of deaths per tonne of coal mined was, on average, eighty-eight times higher than in the US.[28]

Damage to the environment is also severe. Toxins, such as arsenic and mercury, released in coal mining, leak into nearby rivers. Acid rain, produced from sulphurous emissions and now affecting almost a third of the land mass, damages both crops and natural ecosystems.

Perhaps the most deadly pollutant of all is colourless, odourless carbon dioxide, the chief culprit of anthropogenic climate change. Of all the fossil fuels, coal produces the greatest quantities of carbon dioxide per unit of energy it produces. China's production of carbon dioxide per capita is less than a third of the US's, but China's vast and quickly growing population makes it the largest overall producer of carbon dioxide in the world.[29]

China's use of coal has fuelled unparalleled growth and reductions in poverty over the last three decades. And as China continues to grow, it will need more energy. By 2035, China is projected to consume almost as much energy as Europe and the US combined. Coal is the obvious fuel choice to meet this growth since it is cheap and abundant. Its proven reserves alone are expected to last for more than seventy years at current rates of consumption.

Herein lies China's dilemma: how can it continue its rapid economic growth and reductions in poverty, but do so sustainably? The scale of the challenge is unprecedented: China is home to one-fifth of the world's population and in 2010 its use of energy surpassed that of the US. Zhou Shengxian, China's Environment Minister, warned in 2011: 'In China's thousands of years of civilisation, the conflict between humanity and nature has never been as serious as it is today.'[30]

Changing China

In recent years, pollution has become a political topic in China. Internal protests against polluting industries have become more frequent, while China faces increasing international pressure to enforce environmental regulations. The 2008 Olympics made clear that China wanted to be regarded as a responsible global stakeholder. Months before the Games, many industries near Beijing were shut down in an attempt to clear its polluted skies.

However, China's hand is not being forced. It knows that clean energy production is necessary if its social and economic growth is to become sustainable. Air pollution affects food production and the availability of clean water. The deterioration of resources and the natural environment are already bottlenecks to development. The consequences of climate change would be even more serious. China now realises that it is in its interest to take action against pollution.

And action is taking place. China's mining operations are improving, although they remain dangerous. Industrial smokestacks are gradually clearing. Between 1995 and 2004, the air pollution emitted for each unit of GDP dropped by around 40 per cent. Coal is now banned from use in heating and cooking in Beijing and Shanghai and average levels of particulate matter across the country are falling.

In its Five-Year Plan of 2011, China's objective changed from maximising growth to balancing growth with social harmony and environmental sustainability. Its Premier at the time, Wen Jiabao, talked of shifting its model of economic development to 'achieve green, low-carbon and sustainable development'.[31] For the first time, the performance of local officials might be judged not only by economic output, but also on environmental and social performance.

Coal has brought incredible change to China, as it did to Great Britain three centuries ago, but only now, after more than three decades of unsustainably polluting economic growth, is China beginning to manage this relationship between the beneficial and harmful consequences of its use of carbon.

OIL

From an early age, Henry Ford understood the power of carbon fuels. As a young boy, he built a steam turbine next to the fence of his school. It exploded, setting the fence on fire and slicing open the boy's lip. 'A piece hit Robert Blake in the stomach,' he wrote in his notebook, 'and put him out.'[32]

A constant tinkerer with an obvious aptitude for mechanics, Ford's steam turbine was just one of his childhood inventions. On the farm where he grew up he was always finding new ways to ease the drudgery of rural life. 'There was too much work on the place,' he later recalled. 'Even when very young I suspected that much might be done in a better way.'[33]

The realisation of just how that might be done came when he was twelve years old, while travelling by horse and cart to Detroit with his father. Ahead of them Ford saw a cart drawn very slowly by a steam engine fuelled by coal. That scene made a lasting impression on him; forty-seven years later he recalled that engine 'as though [he] had seen it yesterday'.[34] The cart showed Ford how coal could enhance human muscle and mobility. It would take almost two decades more before Ford would realise the potential that oil had as an engine fuel.

On Christmas Eve 1893, he brought his experimental engine into the kitchen of his home where his wife, Clara, was preparing dinner for the following day. His rudimentary invention did not have its own ignition system and so he wanted to create a spark using the kitchen's electricity supply. He told Clara to feed oil slowly into the engine, while he turned the flywheel, sucking a mixture of hydrocarbons and air into the engine cylinder. He created a spark, flames appeared, the sink shook and then the engine began to turn. This instilled in him a lifelong interest in the gasoline engine and its application in the automobile.

Nowadays, we take for granted the freedom we are given by the motor car; we forget how much effort it had previously taken to travel only a few miles. The nineteenth century was a world in which pack horses would struggle along rough trails and makeshift roads, 'where for long seasons every family was practically farm bound, and where isolation wrapped men and women in cobwebs'.[35] By the time the car was developed at the end of that century, iron railroads had forged connections between America's major towns and cities. Goods and people could move further and faster than ever before. Nonetheless, large swathes of the countryside, in which most of the population still lived, remained cut off from the centres of the US's prodigious industrial growth.

The car became the new tool with which to conquer this 'battle with distance'.[36] And only oil, with more carbon-hydrogen bonds than coal, and so a greater energy density, could generate the force needed to power the automobile.[37] The first ones were toys for the rich, custom-built by hand and very expensive, but Ford wanted the everyday person, the farmers and the mechanics, to drive. By pioneering mass production he made automobiles in such large quantities that economies of scale allowed him to make them affordable to a much wider market. The Model T, first produced in 1908, was the culmination of this plan: a lightweight and reliable, yet ultimately very cheap car to meet the growing consumer aspirations of the American public.[38] Fifteen million Model Ts were manufactured and, in 1920, they constituted almost half of all the automobiles in the US.

New employment opportunities, better education and medical care were all now available to the owner of a car. As cities grew, they were shaped around the car in the formation of highways and suburbs. More than a practicality, the car quickly became a status symbol in emerging consumer society. And as a symbol of freedom and prosperity, it became an essential component of the American Dream.

Today the American Dream has gone global. In China's cities, the once impenetrable streams of bicycles have been replaced by cars and the wide thoroughfares now struggle to cope with the ever increasing mass of vehicles. Before the city imposed restrictions in 2011, almost 1,000 cars and trucks were being added every day in Beijing alone.

This global growth drove the sharply rising demand for oil through the twentieth century. However, long before the invention of the car, oil was mostly used for much simpler ends: illumination and lubrication.

Rock oil

Petroleum, the 'black juice' that 'flows from rocks', was first documented by Georgius Agricola in the sixteenth century.[39] In *De re metallica*, his seminal work on mining and metallurgy, he describes how liquid bitumen, found floating in springs, streams and rivers, could be collected in buckets or, when found in smaller quantities, collected with goose wings, linen strips and shreds of reeds.[40] One woodcut in *De re metallica* illustrates a man patiently collecting his haul in a bucket.[41] In Agricola's day, oil was regarded as very inferior to the metallic ores from which iron, gold and silver were extracted. There was little demand for it, but this started to change in the middle of the nineteenth century. The Industrial Revolution created a growing and increasingly wealthy population, which wanted a bright and clean artificial light source. 'Rock oil', as crude oil was then known, was just that, but it was in short supply and therefore expensive. Seeing the potential for profit, wildcat explorers began to search for new and bigger sources of this oil.

In August 1859, Edwin L. Drake, known as 'The Colonel', struck oil 20 metres below the surface on a farm in Pennsylvania. Attaching a simple hand pump to the well, he amazed onlookers by easily pumping oil from the ground. His find sparked a rush to Oil Creek that increased oil production from practically nothing, to three million barrels a year only three years later. Today global oil production stands at thirty billion barrels a year, a ten thousand-fold increase in only 150 years.

Finding that amount of oil has been and continues to be an extraordinary challenge. Explorers need to decide where to drill and how to develop the oil if found. They must do so while making a return on their investment and satisfying the desires of a host government and the needs of affected communities. All this must also be done safely, without damage to the natural environment and with some extraordinary technology.

Randomly drilling wells to find oil would achieve little. It would be like trying to find needles in a haystack. There are always clues that guide explorers to the areas more likely to be winners. An oilfield has certain essential characteristics. First, there needs to be a source for the oil. This source is the remains of plants and animals laid down millions of years ago, which have been subjected to the right pressure and temperature to form oil. Flying over the forests in the centre of Trinidad, I saw lakes of inky black oil which had bubbled up from just this sort of source. The La Brea tar pits in California were formed in a similar way. Both were clues to the presence of other oilfields. Second, the oil needs to get from the source into an overlying structure which can trap it. This often has a dome-like shape (an 'anticline') that sometimes expresses itself on the surface. Anticlines can be seen from the air in the Zagros foothills of Iran, my childhood home. These are the site of some of the greatest oilfields in the world. Third, the trap needs to be sealed by an impermeable rock. If the seal is breached the oil escapes. One of the most famous and most expensive wells, called Mukluk, that turned out be unsuccessful (a 'dry hole'), was drilled in Alaska. For years, BP's explorers were convinced that it was going to be a guaranteed success. It failed because the seal had been breached and the oil had seeped away. Finally, the structure needs to be filled with a sedimentary rock that can contain the oil in its pores (the so-called porosity) and let it flow. The ease with which the rock allows the oil to flow is called the permeability. If all these things come together then there is an oil reservoir.

Some of these characteristics can be identified by analysing how the geology of a place came together. For example, ancient river deltas often have good porous and permeable sand. In modern times, remote sensing is used to 'see' the oil reservoir many kilometres down, below many layers of rocks. The most important technique is seismic surveying in which pressure waves are used to 'bounce' off the deep rock strata. The way in which these seismic waves are transmitted and received needs very careful design. Complex algorithms and gigantic computer power is needed to analyse the collected data. There is not much signal and plenty of noise so the computer analysis is critical. In the best of circumstances the shape of the structure can be identified, the seal can be seen, the rock in the reservoir can be characterised and even the oil in the rocks gives off a particular signature. In more normal circumstances some but not all of these things can be identified. All of this is much harder if the reservoir is overlain by a thick salt layer, as is the case, for example, offshore in Brazil, Angola and the US Gulf of Mexico. Vibrational waves travel so fast through the salt that the layer masks the geological formations that lie underneath. The sudden change in the speed of the seismic waves, as they move from sediment into salt, causes the waves to be refracted and reflected so that the salt sheet acts like a mirror. Only recently have computer algorithms improved to allow these signals to be untangled and sub-salt reservoirs to be seen with a degree of accuracy.

Developing the oil

After all this is done, a well must be drilled to see if oil is there. Sometimes there is success but often not. In recent years the technologies of remote sensing have reduced the chances of failure but success is never guaranteed; Mukluk is a constant and powerful reminder.[42]

Once oil is found, the field then needs to be developed in such a way as to provide a return for the investor and rent for the owner of the subsurface rights, usually a government. But oilfields which are easy to develop are now mostly in production. What is possible to develop continues to change. In the first decades of the twentieth century companies began to move from developing oilfields on land to going out into the water and drilling wells in lakes, in Venezuela, Texas and Louisiana. Going out into the sea was more difficult because the water was deeper, the winds stronger and the waves bigger.

The most extraordinary offshore development I have ever seen is in the Caspian Sea off Baku in Azerbaijan. Oil slicks had been reported by local ships captains and, further out, rocky outcrops were coated in a black oily sheen. In 1947, the first ramshackle drilling platform was erected on the oily Caspian Sea rocks. More and more platforms followed, connected by makeshift wooden bridges. Boulders were shipped out from the mainland to build artificial islands. By 1955, Oily Rocks, as the stilted town came to be known, had become Azerbaijan's largest producer of oil, exporting more than fourteen million barrels each year. Five-storey apartment blocks, shops and hotels rose out of the waters to house and sustain the increasing number of men working there. But the rest of the Caspian Sea's rich oil reserves lay east of Oily Rocks, in deeper water. To tap these fields, sometimes as much as six kilometres below the surface, would demand more than just platforms propped up on stilts at the edges of the sea.

I first visited Baku and Oily Rocks in 1990, following the collapse of the Soviet Union, as BP began to negotiate a venture to explore that deeper part of the Caspian Sea.[43] The state of the town was shocking: bad practice was rife and everything seemed to be leaking. The characters propping up the bars could have been taken out of a scene from *Star Wars*. Trapped behind the Iron Curtain, technology had gone backwards at Oily Rocks along with any hope of accessing the Caspian Sea's deepest reserves. But not all hope was lost. Within the first few years of the new millennium, technology imported from the West was used to develop the super-giant Chirag-Azeri-Gunashli complex.

Developing oilfields in deeper and deeper water required an immense investment in technology. An example is the Thunder Horse field, developed with a monstrous floating structure, which we first met in the previous chapter, 'Iron'. The field was beneath 2,000 metres of water, very different from the 200 metres in the Caspian or 20 metres at Oily Rocks. The capital cost is very high but, every year, technology improves and the cost of producing a barrel of oil decreases.[44] And more can be achieved: developments in even deeper water, of course, but also extracting more of the oil that is in a reservoir. Today, typically 60 per cent or more of the oil is left behind after an oilfield stops producing. The reason for that lies in the economics; extracting more oil becomes increasingly costly and therefore unprofitable. And that cost is being challenged by technology.

Keeping the oil flowing

Henry Darcy, a nineteenth-century French engineer working in Dijon, was a careful observer. He watched water go through the different types of rock at the bottom of public fountains and wondered what controlled its speed. Soon he came up with an equation that described the rate of flow of a fluid through permeable rocks. It is called Darcy's law and the measure of permeability is called the 'darcy' in his honour.[45] The law gives us a way of explaining four different ways in which the flow of oil out of a reservoir can be improved, known collectively as enhanced oil recovery (EOR). First, if the natural pressure of the reservoir is too low to get the oil to the surface, you can increase it by injecting other fluids, such as water, natural gas, nitrogen or carbon dioxide.[46] This is often the first and simplest method of improving the recovery of oil. Second, you can expose more of the reservoir to the well bore by, for example, drilling horizontally along the rock strata. Third, you can make the oil less viscous or prone to staying in the spaces between rocks (the oil is held there by a force called surface tension). One way to do this is to pump in fluids, particularly liquefied carbon dioxide, so that it mixes with the oil. Another way is to heat the oil. This is necessary for so-called heavy oil found in Canadian and Venezuelan tar sands.[47]

Finally, oil recovery can be improved by increasing the permeability of the oil-bearing rock. This is the oldest method of EOR, and one used since the very earliest days of the industry.

In 1865, Colonel Edward Roberts formed the Roberts Petroleum Torpedo Company. He had fought in the American Civil War three years earlier and had observed artillery rounds fired by the Confederate army exploding in dugouts in the battlefield. He thought that a similar blast could improve production from oil wells by fracturing the oil-bearing rock. By filling thin metal tubes with gunpowder, lowering them into a well and igniting them, Roberts was able to create a surge in production from a well, albeit often only briefly. Later nitroglycerine became the preferred explosive medium. Unfortunately it would often detonate by accident, killing and maiming those nearby.[48]

These blunt methods became obsolete with the development of hydraulic fracturing in the middle of the twentieth century. By forcing fluids (often a mix of water, chemicals and sand) into rock formations in which oil or gas is trapped, hydraulic fracturing greatly increases the effective permeability of the rock so that more oil and gas can move to the surface.

There is a great deal of potential for EOR today. In some fields, as much as 80 per cent of the oil is left behind. In others, no production can be had without hydraulic fracturing; this is the case in the so-called 'shale gas and tight oil' developments in the US.

For some time it has usually still been cheaper to find and produce from new oilfields than to try and squeeze more out of ageing ones. In recent years, however, as the price of oil has risen, the potential for EOR has grown rapidly: in the five years up to 2009 the market for EOR was estimated to have increased by twenty times.

The end of 'peak oil'

The amount of oil which is developed is not just driven by technology; it is also determined by future expectations of oil prices. These are, of course, governed by supply and demand. However, the presence of the

Organisation of Petroleum Exporting Countries (OPEC), a cartel which controls around 40 per cent of global oil production, means that supply is often managed to achieve particular price levels. In very simple terms, oil prices, according to OPEC, should not be so low as to cause damage to their economies or cause domestic dissent or revolution; but they should not be so high as to dent demand or encourage too much supply from outside OPEC. Prices have varied broadly between these limits for many years.

All mineral resources are, of course, ultimately finite, and as a result there has always been a concern that the world may be about to run out of oil. In 1956, the American geologist Marion King Hubbert concluded that we would reach a point of maximum oil production, known as 'peak oil'. Using estimates of future consumption and reserves, he predicted that this would occur sometime around the year 2000.[49] But the year 2000 came and went without 'peak oil' happening. Indeed, it is not even on the horizon. This is because we are increasing the world's oil reserves faster than we are using them. And that is mostly down to technology; we find oil in new places and invent new ways of recovering more of what has already been discovered. Increasing the recovery of our existing oil reserves by only 1 per cent would increase them by around ninety billion barrels, equivalent to about three years of global demand. As technologies improve, the percentage of recoverable oil keeps increasing and so supplies continue to get larger. I see no reason why this will cease soon and as a practical matter we probably will not run out of oil. We are more likely to have stopped using it long before we run out. As Sheikh Yamani, Saudi's former oil minister said in the 1970s: 'The Stone Age didn't end because we ran out of stones.'[50]

The financial and technological effort needed to deliver a barrel of oil is extraordinary. To make the oil into petrol takes even more work. It needs to be refined into just those types of hydrocarbons that an engine needs. Even at $4 a gallon, gasoline seems cheap after all this work. It is, after all, cheaper than most bottled mineral water served in chic restaurants in New York City, London or, indeed, Florence, where, a few years ago, I was stunned when offered a small bottle of Tennessee mineral water that cost more than a whole barrel of oil. And bottling water is certainly not as risky as producing oil.

Accidents keep happening

24 March 1989: the news came through just as I was saying farewell to the BP team on Alaska's North Slope. The *Exxon Valdez* oil tanker had run aground on Bligh Reef in Prince William Sound. The ship had been sailing outside the normal sea lane to avoid icebergs, but in doing so had run aground on the rocky seabed. We were about to fly back to Anchorage and so diverted the plane to fly over the tanker. As the *Exxon Valdez* came into view ahead of us, we could see her side split open and oil seeping out to form a black smudge around her. Soon, this small pool of oil would spread to cover an area of 30,000 square kilometres.

This was by no means the first or the largest oil tanker spill. As I stared out of the aeroplane window at the *Exxon Valdez* below, I thought of *Amoco Cadiz*, another oil tanker that ran aground off the coast of France in 1978, and of the *Atlantic Empress*, which collided with another ship off Trinidad and Tobago in 1979. Between these accidents 500,000 tonnes of oil was spilt, compared to 39,000 tonnes from the *Exxon Valdez*. However, the Alaskan accident was the first time so much oil had been released into an environmentally important ecosystem.[51] Storms pushed the thick black oil on to the rocky beaches, covering 2,000 kilometres of pristine coastline and resulting in the deaths of 250,000 seabirds. Many otters and seals were also killed and the herring population, on which many local people depended for their living, was virtually wiped out. Images of the oil slick set against the white backdrop of the Alaskan mountains filled the front pages of newspapers around the world. Seabirds, feathers black and tangled, were held up to the camera lens, an emotional reminder of the dangers of oil extraction.

Oil is a liquid and so spills are hard to contain, while the flammable nature of oil, and the gas found with it, can create a serious danger to human and animal life. On 6 July 1988, while I was working for BP in Ohio, the news came through that there had been a serious accident on Occidental Petroleum's *Piper Alpha* platform in the North Sea oilfield. A large number of gas explosions had engulfed the platform in fire. Workers who had not been killed in the blasts were left sheltering in the increasingly smoke-filled accommodation block. Others took their chances and leapt 60 metres into the water below. Of the 226 people on the platform at the time of the accident, 165 died. The incident sent shock waves through the oil industry: everyone realised that it could just as easily have happened to them. Big advances were made in process safety in the wake of the *Piper Alpha* disaster, but they were not enough to prevent the saddest and probably worst day of my working life in March 2005, when an explosion occurred at BP's Texas City refinery, killing fifteen and injuring more than 170.[52]

Most recently and most dramatically, in April 2010 a bubble of methane gas escaped from the Macondo oil well in the deep waters of the US Gulf of Mexico. The blowout preventer had failed; as a result, gas rose up the last sections of drill pipe and reached BP's *Deepwater Horizon* drilling rig. The gas ignited, causing an explosion that killed eleven people. Macondo became unplugged. Oil began to seep into the ocean and continued to do so for almost three months. I watched with despair as the disaster unfolded and cameras, more than 1,500 metres underwater, showed the oil leaking into the sea. Macondo was the first real-time industrial disaster; you could switch on the TV and watch the oil seeping out every hour of every day.

There are significant risks involved in our extraction of oil resources. Even with lessons learnt from past incidents and the very best of practices, accidents continue to happen. For example, between 2000 and 2010 there were, on average, just over three large oil spills a year, in which more than 700 tonnes of oil were spilt. The unprecedented scale of the Macondo explosion and oil spill was a sharp reminder that, as we push out the limits of technology, we expose ourselves to new things going wrong. In hindsight, the cause of an explosion or spill seems obvious, with mechanical inadequacies amplifying human error. But neither equipment nor human processes can be made entirely risk-free and safe. And this is something we all should remember as we take oil for granted.

Technology underpins oil's progress but this is merely one side of the story. The oil business is one of the most ruthless and cut-throat in the world. The politics of oil, the characters, companies and countries who struggle to control reserves, determine how this form of carbon contributes to our lives. In that mix, one individual stands out as having the single greatest impact on the development of the oil industry.

Oil conveys power

John D. Rockefeller looked upon the early wildcat oil prospectors with disdain. On a visit to Oil Creek, near Colonel Drake's original oil strike in Titusville, Pennsylvania, he was appalled by their loose morals. Thousands had swarmed to the area in the hope of striking lucky. They brought with them disorder, verging on chaos, sinking wells wherever they could buy land. Where, when and how much oil would be discovered was anyone's guess and prices fluctuated wildly. Soon, so many men were producing oil that prices began to fall sharply because of overproduction. Rockefeller, then the owner of a Cleveland refinery, decided that something had to be done. In January 1870 five men, led by Rockefeller and his partner Henry Flagler, founded the Standard Oil Company. His goal was simple: to combine the dominant oil companies of Ohio and so reduce excess capacity and take control of the price fluctuations that threatened his business.

Rockefeller was as daring as the early oil wildcatters, but he was also rational, calculating and measured. From an early age he appreciated the economies of scale. As a boy, he would buy candy by the pound, before dividing it into smaller portions to sell at a profit to his siblings. Rockefeller recognised not only the profitability of scale, but also the stability that it brought. He integrated his own oil supplies with his own distribution network and created his own infrastructure, at first making his own barrels, and then buying pipelines, tankers and trains. Rockefeller enhanced Standard Oil's competitive position and insulated its overall operations from the actions of others in the market. So big were Rockefeller's operations that he could force discounts from the rail and shipping companies that were not his own, and so undercut competition further.

Rockefeller's business practices were ruthless and his tactics were often underhand. If a competitor did not agree to being bought out, Rockefeller would give them 'a good sweating', cutting the price of oil products in that particular market and forcing them to operate at a loss.[53] With no choice but to sell, Rockefeller would get the company at a discount. One by one he squeezed his competitors out of the game. In Cleveland, Standard Oil managed to acquire and shut down twenty-two out of its twenty-six competitors in less than two months. By 1879 it controlled 90 per cent of America's refining capacity and much of the US oil pipeline and transportation network.

The public regarded Standard Oil as too powerful, very devious and extremely ruthless. The company's operations were entirely opaque and they seemed accountable to no one. Standard Oil failed to realise the full scale of public opposition towards them. In 1888, one senior executive wrote to Rockefeller, in a manner sometimes imitated in corporate life even today: 'I think this anti-Trust fever is a craze, which we should meet in a very dignified way & parry every question with answers which while perfectly true are evasive of bottom facts.'[54] In 1911, to restore competition and reduce its power, Standard Oil was broken up under the Sherman Antitrust Act.[55]

Rockefeller's reputation is therefore that of a 'robber baron' who sought only personal gain with no regard for the hurt he caused his workers, their families and others in the industry. Most famously, the American journalist Ida Tarbell wrote that it was 'doubtful if there has been a time since 1872 when he has run a race with a competitor and started fair'.[56] Tarbell believed Rockefeller to be 'the victim of a money-passion which blinds him to every other consideration in life'.[57] In this respect Rockefeller was truly successful: he became the world's first nominal billionaire in 1916.[58]

However, Rockefeller's strong religious beliefs led him to behave in an apparently contradictory manner: he decided to give much of his money away. In his philanthropic gestures, Rockefeller was remarkably forward-looking. In 1882, he gave money to a black women's school, when higher education for both women and blacks was frowned upon by many. He believed that philanthropy at the time was 'conducted upon more or less haphazard principles', and he sought to make it more effective.[59] He would take a business-like approach to each potential beneficiary and make sure that he never gave them so much money that they would become dependent on his gifts. He wanted people to be able to help themselves, writing that 'instead of giving alms to beggars, if anything can be done to remove the cause which lead to the existence of beggars, then something deeper and broader and more worthwhile will have been accomplished'.[60] He was in the vanguard of attitudes to 'self-improvement'.

Rockefeller's behaviour is, in some ways, reminiscent of today's Russian oligarchs. They have accumulated wealth on a similar scale, many having started from poor backgrounds just like Rockefeller. They have also made their money by, at times, foul and unfair means.[61] And like Rockefeller, some of the oligarchs have now matured to a point where they want to give some, or all, of what they have earned back to society.[62] But this did not seem to be the case when I first encountered them during Russia's 'Wild East' capitalism of the 1990s.

Russian risks and rewards of oil

18 November 1997: watched over by the British Prime Minister, Tony Blair, and Russia's First Deputy Energy Minister, Viktor Ott, I signed a contract worth about US $600 million with Vladimir Potanin, one of Russia's most powerful businessmen.[63] The deal was for a 10 per cent share in the oil and gas business Sidanco and it was BP's first step into the country. Russia had improved since I had first visited in April 1990, but the country was still a wild and lawless place and none of us could forecast in what direction it was going to move. By signing at 10 Downing Street, London, and with Blair and Ott acting as our 'godparents', BP had hoped to protect itself against the more underhand of dealings.

Getting to this stage had not been easy. In the chaos that surrounded the collapse of the Soviet Union, output from the oil and metal companies had fallen dramatically. The cash-strapped government had then sought to sell them off to businesses or individuals in return for loans. As a result, seven individuals acquired many of the state's assets at what turned out to be a knock-down price, mostly through the notorious 'loans for shares' scheme, which came to be known as the 'sale of the century'.[64] The seven individuals became known as the oligarchs; they were the ones resolute enough, and in some ways lucky enough, to come away with an extraordinary prize. Potanin was one of them and, at one stage, his empire controlled about 10 per cent of Russia's GDP.

In summer 1998, BP began to notice that Sidanco's oilfields, held in many smaller subsidiaries, were gradually disappearing. A new bankruptcy law passed earlier that year was being used to acquire the assets at a heavily discounted price. A general manager of one of Sidanco's subsidiaries would issue short-term debt, which would then be bought up by a third party. On maturity, repayment would be demanded, but the general manager would decline to pay the debt. The third party would then take the subsidiary to trial in bankruptcy court, far from Moscow and without telling BP. It would always win and get the subsidiary at a knock-down price; the general manager would be rewarded for efforts in the scheme.

In November 1999, BP learnt that Sidanco had lost its biggest asset, Chernogorneft, which was responsible for three-quarters of Sidanco's production. The field had been auctioned off through the same farcical bankruptcy process to the Tyumen Oil Company (TNK). By now BP had lost most of its initial investment, but if it had allowed itself to be pushed out of Russia it would almost certainly never be able to go back. Russia was too important an oil province for BP to leave and so it had to play TNK at its own game. TNK had borrowed a lot of money to buy up the Sidanco debt and BP was able to trace these loans to a number of Western banks. BP told the banks that the credit they had supplied was being used to support corruption. Slowly the loans dried up. As they did, Mikhail Friedman, the major shareholder of TNK got in touch with BP.

It turned out that, as is usual in conflicts, the 'theft' of Sidanco's assets was not clear-cut. Friedman was pursuing what he considered to be his legitimate rights as an early shareholder who had been unfairly bought out by Potanin for a fraction of the asset value. Potanin had been weakened by the 1998 financial crash in Russia and so Friedman had decided to take advantage of him. Eventually BP reached an agreement with Friedman: the company would get its assets back, and acquire half of TNK for an investment of $8 billion. BP seemed to be firmly established in Russia. As 2012 drew to a close, though, BP and its partners were ready to sell TNK-BP for a sizeable profit to a state-controlled oil company, Rosneft. In less than two decades, the 'sale of the century' had been reversed.

The oligarchs had become more successful and more powerful. Over time, they wanted to protect their own wealth both from the reach of the government, which could apply the law to its advantage, and from simple theft. In this way they were like the nineteenth-century American robber barons. They used their political influence to keep as much power as possible for as long as possible. In history, the development of oil tends to concentrate power and wealth in the hands of only a few. Even after the 'dragon' that was Standard Oil had been slain in 1911, the power of oil continued to remain in the hands of a few.[65]

From Standard Oil to the Seven Sisters

For the first half of the twentieth century the so-called Seven Sisters, consisting of the largest fragments of Standard Oil, along with rival US companies Texaco and Gulf and European companies British Petroleum and Royal Dutch Shell, controlled production from the majority of the world's oil reserves.[66] Working as a cartel, they respected each other's markets and were able to eliminate other competitors, much as Rockefeller had done in the US.

The Seven Sisters also received support from their domestic governments. In 1914, Winston Churchill, then the First Lord of the Admiralty, purchased a controlling stake in the Anglo-Iranian Oil Company for the British government to ensure a secure oil supply for her navy. Similarly, the US government asked its oil companies to participate in arrangements in the Middle East that were contrary to the same laws which had been used to break up Standard Oil.

The initial success of British Petroleum and Royal Dutch Shell depended on their oil discoveries in the British and Dutch empires. Although these empires were gradually dissolved after the Second World War, colonial attitudes remained and the Seven Sisters continued to exert considerable power over oil-exporting countries until the 1940s.

The balance of power began to shift from the oil-consuming countries to the oil-producing countries when Venezuela negotiated the first 'fifty-fifty' deal in 1943. Under the new petroleum law, government oil revenues would equal the profits after taxes and royalties in Venezuela of any oil company operating there. The fifty-fifty principle soon spread to the Middle East and became standard practice across much of the global oil industry. Many oil-exporting countries wanted more: the oil belonged to them and so, they believed, they should receive the lion's share of profits.

The 'fifty-fifty' principle was eventually broken in 1957 when Enrico Mattei, president of ENI, the Italian multinational oil and gas company, conceded to an unprecedented 25:75 split between ENI and Iran. Italy wanted a share in the new giant Middle Eastern oilfields, known as 'elephants', that were discovered in the 1950s. Mattei was derisive about the close ties between large international oil companies and wanted to weaken their stronghold on the world's oil supplies. In his aggressive pursuit for reserves, he struck a deal that no other oil company was prepared to agree.

The weakening of the Seven Sisters along with the sudden discovery of vast new oil reserves provided opportunities for new oil companies to get involved in the game. The growth of new reserves also resulted in a surplus of oil. Unable to compete with low Soviet oil prices, oil companies began to cut the price they were prepared to pay for their oil (the 'posted price'), but this also cut the national revenue of oil-exporting countries.

In response, in September 1960 five major oil-producing nations formed OPEC. In doing so they hoped to take control over the setting of the international oil price. However, supplies were abundant and oil companies, which controlled the market outlets, treated the oil-exporting countries with disdain. Rivalries inside OPEC made price setting even harder and so the price of oil fell throughout the decade.[67] Not only had OPEC failed to raise the price, they had failed even to defend it.

In the 1970s demand caught up with supply. Realising the power they now held, in 1973 OPEC unleashed the 'oil weapon'.

Anxiety about oil supplies

On 6 October 1973 a coalition of Arab states led by Egypt and Syria launched a surprise attack on Israel, starting the Yom Kippur War, known as such for commencing on the Israeli holy day of Yom Kippur. By choosing for the attack a day of religious fasting and rest, the Arab forces hoped to catch Israel when it was least prepared. Although Israeli armed forces were strong, they had miscalculated how long their supplies would last. As they ran out of equipment against the Soviet-supported Arab forces, Israel asked for help from the US, who then provided it.

In response to US support for Israel, Arab members of OPEC cut production by 5 per cent from the September level. They then announced a further 5 per cent cut each month so as to increase pressure on the US. With gradually diminishing oil supplies, the US economy was being stifled.

The war continued and, later in October, President Nixon announced a $2.2 billion military aid package for Israel. Saudi Arabia, the world's largest oil exporter at the time, retaliated by announcing a complete embargo of oil shipments to the US. Other Arab states soon followed suit.

Prices rocketed to over $50 a barrel of oil in today's money, the highest since the wildcat days of the oil industry at the end of the nineteenth century. The price increases were worsened by OPEC's announcement in the previous week that it was taking complete authority over the setting of the posted global oil price, raising it by 70 per cent. These events marked the beginning of our modern anxiety over oil supplies.

At the time I was living in New York and working on Alaskan oil developments. On OPEC's announcement, the city was thrown into a state of turmoil. Queues for gasoline stations ran for blocks down the street. People would drive from station to station searching for fuel. Even with their tank nearly full, they would wait for hours for a top-up; who knew if there would be any gasoline tomorrow? The embargo galvanised America into action: in January 1974, we received the long-awaited authorisation to build the Trans-Alaska Pipeline to run between our super-giant oil field in Prudhoe Bay in the north of Alaska and the Valdez terminal in the south.[68] It would eventually start production in 1976 and at its peak in 1988 supply over two million barrels of oil per day, equivalent to almost 12 per cent of the US's demand for crude oil.[69]

The oil crisis of 1973 marked a new relationship between those who produced oil and those who consumed it. No longer was oil a plentiful resource, the supply of which could always be assured. It was now a political weapon, a vital strategic interest and the cause of frequent crises in the global economy. The next of these crises was to emerge in the Iranian Revolution of 1978. Iran then provided 20 per cent of the world's oil, and so the global supply was disrupted and the price again rose.

Iran's rich oil reserves were a cause of the Iran–Iraq war of the 1980s and also Iraq's invasion of Kuwait in 1990. Wary of neighbouring Iran's great oil wealth, Saddam Hussein wanted to increase Iraq's power, and one way of doing that was to increase his oil reserves. As Iraq invaded Kuwait in August 1990, oil wells were abandoned, disrupting global oil supply once again. Several months later, Iraqi forces fled following a US-led military attack, but not before they had set Kuwait's oil wells alight, stopping production for months afterwards. We all vividly remember images of the thick black smoke rising from Kuwait's burning oil wells. BP had developed the Kuwait oilfields and so, being the then sole possessor of information on the fields, cooperated in the clean-up operations.[70] Later that year, I visited Kuwait and was shown around the fields. The fires had only just been put out and all that remained was a mass of twisted, molten equipment. The entire desert was black; it looked as though it had been asphalted.

Unstable politics, increasing demand and OPEC's tightening grip on oil supplies led to high prices through the 1970s. However, high prices also stimulate greater investment in exploration and production. This is just what happened and by the middle of the 1980s new oil supplies, combined with an economic downturn, caused the price of oil to slump to very low levels, which persisted through the 1990s. This presented oil companies with a new challenge altogether.

To survive with the low oil prices of the 1990s there was a need for consolidation in the industry so as to benefit from economies of scale. BP was too small. If BP did not acquire another company it would surely have been taken over by another. I began discussions with Larry Fuller, chairman and CEO of Amoco, and on in August 1998 I announced that BP would merge with Amoco, triggering a wave of mergers across the industry: Exxon merged with Mobil, Chevron with Texaco, Conoco with Phillips and Total with Fina and Elf.[71] Between 1998 and 2002 the industry went through the most significant reshaping since 1911 and the break-up of the Standard Oil Trust. The biggest deal, the merger of Exxon and Mobil, brought together the two largest companies to have emerged from its 1911 dissolution. Scale brought efficiency, security and clout, enabling these newly created super-majors to compete with states and governments and take on bigger challenges of greater technological complexity. The risks of oil extraction could be spread over a greater number of barrels of oil.

In the 1990s, oil prices were low and national oil companies (NOCs) lacked the confidence to compete. The size of the super-majors, combined with their expertise, made them valuable partners for producer states. But today the growing size of NOCs combined with high oil prices has once again shifted the balance of power back towards producer states. Around the world, producer states have reduced the share of rents given to the super-majors. In Bolivia, the government has seized oilfields outright; in Venezuela the government has rewritten contracts to give the national oil company control; and, most recently, in Argentina a controlling stake in the former state-owned energy company YPF was seized from Spain's Repsol.

The relationship between the super-majors and oil-producing nations is often tenuous: one side answers to its shareholders, the other to its citizens. Both sides want to extract as much natural wealth as possible; the question of who gets what remains one of power.

The problem of Ricardo's rents

It was 1999 and I arrived at Claridge's hotel in London where I was to meet Hugo Chávez for the first time since he had become president of Venezuela the previous December. I was hopeful. Chávez had not given details about his plans for his country's oil industry, but he had said that he regarded continued foreign investment to be of vital importance.

Since 1993, under the presidential leadership of Carlos Andrés Pérez and later Rafael Caldera, international oil companies were being invited back into the country as part of the opening of the economy, *la apertura*.[72] The new head of the state oil company Petróleos de Venezuela SA (PDVSA), Luis Giusti, played a central role in this, as, indeed, had BP's general manger on the ground, Peter Hill. During the years of 'la apertura', there was a constant flow of international talent and capital into the country. Billions of dollars had poured in; production from Venezuela's declining oilfields was being increased. There was a great sense of possibility within PDVSA and international oil companies alike.

Facing Chávez in the hotel room, it soon became clear to me that this was all about to change. A few minutes into the meeting he started on a speech about the 'evils of foreign oil companies.'[73] To Chávez we were the new face of those who had, for centuries, pillaged the natural riches of South America; to rest of the world, Chávez was the new face of resource nationalism.

The conflict between private oil companies and oil-producing states centres on the problem of rents, an idea first formulated by David Ricardo at the beginning of the nineteen century.[74] Ricardo's Law of Rent assigns a value to land based purely on its intrinsic 'natural bounty'. Venezuela was traditionally an agricultural nation, dependent on the production of cocoa, coffee and sugar. The low value of these goods could only support a small and poor population; the land had little intrinsic value. The discovery of oil in the early twentieth century created a sudden shift in the Venezuelan economy. The price of oil was high enough so that, after all costs of production and a reasonable return to investors, a surplus was left. This surplus is defined as rent, the inherent value of the land.

Who should receive this rent and how much should they receive? Is it the oil companies that extract the oil, or the state that owns the land? Both hold a claim. Oil companies take great risks in exploring and developing production in foreign countries – in Alaska, for example, BP nearly walked away with huge losses following a fruitless ten-year search. On the other hand, the land and its 'natural bounty' are the property of the state. Any state will want to get as big a share as possible. Companies are invited in to find and develop reserves and, in a perfect world, the rules under which they operate would remain unchanged. However, experience shows that, when oil starts flowing, the state squeezes. The investment is sunk and oil companies are defenceless. But there are only a limited number of times the state can get away with these tactics before investment is frightened away. Eventually, companies leave, taking their highly trained people, expertise and investment with them.

The progressive oil industry reforms implemented by Giusti and Pérez in the early 1990s were beginning to reverse the economic decline of the 1980s, but Chávez's revolution put an abrupt stop to this.[75] When Chávez came to power in 1998 he quickly denounced Giusti as 'the devil who had sold the soul of Venezuela to the imperialists'. Now the state wanted all the rent. PDVSA quickly became 'the cash box' of the state. With financial control of the company in the hands of the central government, the state could now do what it wanted. Following a strike against his presidency in 2002, Chávez fired almost half the workforce. Many of his allies, largely military personnel, were put in power at PDVSA.

Since Chávez's election, Venezuelan oil production has fallen by 20 per cent, from 3.5 million to 2.7 million barrels a day. Fortuitously, increasing oil prices worked in Chávez's favour, masking the decline in production. But he had lost valuable expertise and investment, severely harming the long-term development of Venezuela's resources and undermining the contribution that oil could bring to his citizens and the rest of the world.

A similar story can be told about Libya following the expropriations by Colonel Qaddafi in the 1970s. In one extraordinarily naïve moment three decades later, when I made a visit to him in his tent, he asked me why he had lost all the industry expertise he needed and why his oil production was in decline. He admired other African nations and said he wanted to be like them. He realised they had kept their expertise even after nationalising some of the foreign company interests in their countries,[76] but he failed to understand that people go away when there is nothing left in it for them.

Long-term cooperation is needed between private companies and the state if oil development is to remain innovative and sustainable. Globalism and nationalism often seem incompatible. By striking the right balance between these extremes, resource development can benefit everyone. When handled correctly, the relationship is one of constructive tension, producing oil to meet global demands but also providing wealth to the nation state. And this is a lesson to be learnt again and again. Yesterday's Venezuela and Libya are not good examples for today's newly emerging oil states, many of which are in Africa.

A curse or blessing?

'I call petroleum the devil's excrement. It brings trouble ... Look at this locura – waste, corruption, consumption, our public services falling apart. And debt, debt we shall have for years.'[77] What could cause Juan Pablo Pérez Alfonso, one of the founders of OPEC, to denounce oil as the devil's excrement? In 1948, Alfonso had negotiated the first 50:50 split of rents between international oil companies and the state, greatly increasing the state's share of oil revenues. Ever since that time he had seen the harmful impact of oil wealth on the Venezuelan economy and society.

John D. Rockefeller always thought that 'oil is almost like money.'[78] Such easy wealth can breed complacency and, with little to strive for, an oil state's innovation and competition suffer. Time is spent arguing over who gets the money rather than developing the economy. An economy predominantly based on oil is also highly unstable: when the price is high the economy thrives, but when it crashes, so does the economy. Alan Greenspan, one of the greatest central bankers of our time, was well aware of the distorting effects of oil when he told me that he would advise any African leader who had discovered oil to forget about it, and to hide the news. Such is the perceived curse of oil.

The potential for development of a nation that is rich with oil can be further stifled in a corrupt system of governance where the rent never makes it into the hands of its citizens. This can happen during civil conflict or in the aftermath following the departures of colonial rulers. Corrupt leaders can remain wealthy and powerful just by controlling key infrastructure. They need pay no attention to their country's economy or the views of their citizens. This is the 'oil curse'. [79]

In the early 1990s, I was in Angola to discuss BP's possible involvement in the country's offshore oil activity.[80] As I was driven through the war-torn streets of the capital Luanda, the visible poverty made it clear that Angola was falling foul of the oil curse. The civil war began in Angola immediately after Portugal, which had colonised the territory in 1483, handed back power. It started in 1975 and continued, with some interludes, until 2002. The conflict was between two main opposing political movements, the socialist People's Movement for the Liberation of Angola (MPLA) and the anti-communist National Union for the Total Independence of Angola (UNITA).

Angola has great natural wealth in the form of oil and diamonds (a pure form of crystalline carbon).[81] In the initial chaos of the conflict, UNITA captured the best Angolan diamond deposits, while MPLA took control of the oil reserves. Two forms of carbon, both sources of great wealth, were funding opposite sides of a civil war. At first, diamonds and oil were only a means to make war. However, the wealth and power they gave both sides of the conflict also made them a reason to continue fighting. Angola's citizens, caught up in the conflict, fell into deep poverty.

In 1998, the UK-based NGO Global Witness published a damning report on the impact that the illegal diamond trade was having in supporting the conflict in Angola.[82] The report stated: 'The international community … has become complicit with diamond barons', helping UNITA to rearm itself in a war which had caused around half a million deaths during the mid-1990s alone.[83] Diamonds can be used to hide vast wealth in a minute volume and, by avoiding paperwork, traded without trace. UN prohibitions on the sale of these diamonds were being flouted and companies that did so were not being prosecuted. In particular, Global Witness pointed the finger at De Beers, who controlled around 80 per cent of the world's diamond trade.[84]

In the following year, Global Witness published another scathing report aimed at the Angolan oil industry.[85] Oil was, and still is, the main source of revenue for Angola; at the time it represented as much as 90 per cent of the government's income. Global Witness reported that much of the wealth was being siphoned off by corrupt officials, rather than being used to help repair the war-torn country. They suggested that BP take the drastic step of publishing the records of all the contractual payments they had made to the Angolan government. And that was exactly what BP did.

By ensuring transparent disclosure, and the consequent scrutiny of financial flows, BP had hoped to put pressure on the government to use oil money in ways that would benefit the wider Angolan population. In the longer term, it would also limit the taxes BP might have had to pay: if the government used the money wisely they would not come back and ask for more. It was clear that if this idea was to be sustainable, it needed to be made into a standard that applied to all countries. So I went to discuss it with Tony Blair, who agreed to help establish an appropriate institution. In

2002, at the World Summit for Sustainable Development in Johannesburg, Blair announced the creation of the Extractive Industries Transparency Initiative (EITI). The EITI was a mechanism to encourage countries to publish the payments they received from oil, gas and mining companies. Encouraging government cooperation was vital. It does not matter how many companies publish financial records; transparency is only effective if governments also publish spending records for comparison. Only then is scrutiny possible.[86]

The EITI is only a small part of the solution to the oil curse suffered by Venezuela, Angola and others, but the transparency it supports is a necessary starting point. There is plenty of room to do more.[87] For example, the EITI does not cover how exploration and development contracts are awarded. Corruption is often more subtle than the direct siphoning off of state funds. Officials may give contracts to companies that are less qualified if they think they are more likely to give kickbacks.

Transparency initiatives, such as the EITI, are currently voluntary. This must change so that there is nowhere for anyone to hide. Companies must disclose all payments to foreign governments in sufficient detail for citizens to see what is happening to their locally generated revenues.[88] Perhaps then oil will become more of a universal blessing.[89]

Future oil

Finding and developing new oil reserves and recovering more of the old oil reserves will continue to be essential since demand is expected to keep rising. Over the next twenty-five years, it is expected to increase from eighty-eight million barrels a day to one hundred million, mostly as a result of rapid growth of ownership of cars and trucks in Asia. A sufficient diversity of supplies, and a cushion of capacity to produce more oil if needed, will probably reduce our concerns about shortage. Geopolitics will determine the rest: the level of rent taken by governments and the provision of the right incentives to get the appropriate skills so that their oil can be produced.

A remarkable shift has recently occurred that could begin to ease the West's anxiety over oil supplies. What started as a revolution in the production of natural gas quickly spread to the production of oil, and that is where the story of carbon will finish.

NATURAL GAS

Galeota Point, the south-eastern peninsula of the island of Trinidad, is the home of much of the region's diverse wildlife; the tropical ecosystem is an important migratory 'pit stop' for many birds on their way across the Caribbean. It is also the home of some of the island's most significant hydrocarbon reserves, which lie some way offshore.

In May 2004 Patrick Manning, the Prime Minister of Trinidad and Tobago, was showing me the sights of the island by helicopter. The next stop was the vast tar pits in the centre of the island. The inky blotches stood out in the otherwise tropical landscape, a sure sign of the island's rich oil and gas reserves. Trinidad contained some of the jewels that BP had recently acquired in its merger with Amoco, and I was there to ensure BP's growth in the region would continue. BP had made several major gas discoveries in the late 1990s and early 2000s, including the first one in Trinidad's deep water.

Natural gas is mostly comprised of methane molecules. These are made up from one carbon atom connected to four hydrogen atoms. It is the simplest bonding of carbon with hydrogen and one of the most abundant energy sources on earth. Manning wanted to ensure that this latest find of natural wealth would be used for his people's benefit.

Trinidad had clearly been exploited by its former colonial masters. Spanish conquistadors first settled in Trinidad and Tobago in the sixteenth century. At the turn of the eighteenth century, Great Britain took control of the islands until they obtained independence in 1962. Over time, unsustainable agricultural practices had degraded the natural landscape, while slaves were imported from West Africa to harvest the sugar cane crop. As Manning further explained to me, they had exported the sugar produced on the island, but then could not afford to import the boiled sweets that were made from it.

With the discovery of oil in Trinidad in 1886, the economic prospects of the country improved. Disused oil barrels were soon scattered over the island, forming the foundations of the Caribbean steel drum culture. The development of gas on Trinidad followed in the late 1950s and was used on the island for the production of ammonia for fertiliser. In the 1970s, new discoveries led to further growth of the island's gas industry. The government wanted to use gas on the island, rather than export it, and energy-intensive industries became attracted to the area.

The country's dependency on oil and gas, as in Venezuela and Angola, brought its own set of problems. The rise and fall of Trinidad's economy became closely tied to the volatile price of oil, thriving in the seventies when oil prices boomed, but crashing again during the 'oil glut' of the eighties. At the same time, poor performance of the local chemical industries pushed the country into a deep recession.

When I spoke with Patrick Manning in 2004 he was serving his second term as prime minister. His first term was in the early 1990s, when the island's gas industry was growing fast. It soon became clear that gas was by far the island's most plentiful resource, oil having reached its production peak in 1978. With the scars of the oil curse in mind, Trinidad decided to use this gas to develop its energy-intensive industries further, creating new ammonia, steel and methanol plants to supply the large export market in the nearby US. In 2004, with a newly discovered set of gas fields, Manning was once again determined to use the island's natural resources locally. BP helped further by incorporating the local community into its own industry: a gas platform, which went into operation in 2006, was largely constructed in Trinidad using local people and resources.[90] Ensuring economic growth and preserving the natural environment around its activities was BP's mission. If the local people saw harm or did not see a personal benefit from BP's presence, public opposition could quickly have resulted in it being kicked out of the country.

But Trinidad is only a small island, and the size of their large gas reserves holds plenty of potential for export as well as supporting local industry. When I met Manning, he explained his plans to form a pan-Caribbean political union, in which he would trade gas for other Caribbean commodities. Even among Trinidad's Caribbean neighbours there was relatively little demand for gas, and so Trinidad looked further afield to the US. The problem was how to get the gas there cheaply.

Frozen gas

In its gaseous form, natural gas always takes up much more space than an energy equivalent amount of liquid oil.[91] The energy is therefore extremely expensive to transport. A partial solution was found in the middle of the twentieth century with the development of a technology to liquefy the gas into liquefied natural gas (LNG).[92] By compressing the gas and then forcing it through small valves the gas is cooled. This is the so-called Joule Thomson effect.[93] By repeating this process several times, the LNG train, as it is called, cools the gas down to a temperature of minus 162 degrees centigrade so that it turns into a liquid. In doing so, the volume is reduced to one-six-hundredth of the gaseous volume, making shipping an economic possibility. Liquefaction is, however, an expensive process: a typical LNG train consumes about the same amount of power as two million domestic refrigerators, costs billions of dollars and takes years to complete. To be profitable, LNG has to be produced on a huge scale and at the right price.

During the 1980s, LNG from Trinidad could not compete with cheap indigenous gas in the US, where recession had led to an excess supply. So projects to take it to the US were abandoned as was a project to sell gas to Puerto Rico because the buyer would not commit to take the gas for long enough. However, a decade later the US gas surplus had disappeared. There was now a new and profitable export market for LNG. Trinidad's first LNG train, called Atlantic LNG and operated by BP, came into action in March 1999 at Point Fortin in the south-west of the island. The creation of a new export market sparked new exploration in the country and proven gas reserves grew. Between 2002 and 2005, three more LNG trains were added to Atlantic LNG, fuelling a sustained period of economic growth on the island.[94]

LNG enables natural gas to be transported to virtually anywhere in the world that has a terminal capable of receiving the LNG and turning it back into a gas. The development of huge new LNG trains around the world, notably in Qatar and Australia, is transforming a series of regional gas markets into a more global one with the advantage that disruptions to supply, whether caused by political events or natural disasters, can be dampened by quickly sourcing LNG from elsewhere.[95] By 2011, the LNG market had expanded enormously and was a hundred times bigger than it had been forty years earlier. It now accounts for a third of all the gas transported internationally in the world. Natural gas has, relatively rapidly, graduated to become a very valuable source of energy. In China, though, 2,000 years ago, natural gas was viewed as a manifestation of evil.

Pipeline problems

Around 250 BC, when digging began in the salt mines of Sichuan in China, people thought evil spirits seeped up through cracks in the rock. Workers at the pits would suddenly become weak, lie down and die. At other times, the pits would be rocked by great explosions. Each year offerings would be made in the hope of appeasing the evil spirits. Miners soon realised that these deadly occurrences had a more scientific explanation: an invisible, combustible gas that would rise up through cracks in the rock. By AD 100 they were using these fuel streams to boil brine at the mine site, leaving behind residue salt. Pipelines were the next logical step. By AD 200 the Chinese were using bamboo tubes to collect the gas, caking joints with mud and brine to stop gas escaping from this rudimentary piping system. Natural gas could now be transported to boiling houses for cooking and boiling brine on a much larger, and more efficient, scale.

However, salt was still the main preoccupation of the Chinese. In 1835, the Shen Hai well was dug in Zigong to a record-breaking depth of one kilometre. On the way down, at 800 metres, miners hit a reserve of natural gas, but they kept digging. It was salt that they were after; natural gas was merely a useful by-product.

In the past, natural gas has often been regarded as an inconvenience. It is not only found separately, but also found dissolved in oil from which it needs to be separated before the oil can be transported. At the start of the modern oil industry in the late nineteenth century, the gas was simply burnt off as no economic use could be found for it. It was easier to transport solid coal to where energy was needed and then convert it into town gas, the forerunner of natural gas. Not until 1935 did sales of natural gas overtake town gas.

'Gas disposal' was for a long time an important consideration for oil companies since, if gas was not flared, it would inhibit oil production. I remember the roar of the flares burning gas as I played badminton in our garden in Iran. Surprisingly, even today, large amounts of natural gas are flared in Russia since it cannot be used economically. Flaring also continues in other oilfields in inaccessible and hostile environments. Concerns over climate change and simple economics (it may be profitable to use the gas) have reduced this practice. Between 1996 and 2008 the volume of gas flared for each barrel of oil produced decreased by around 30 per cent. There remains great potential to reduce this practice further as around 150 billion cubic metres of gas are still flared each year (around 4 per cent of total global consumption).

The modern gas industry took off in the US in the 1920s with the development of sophisticated networks of pipelines. Advances in metallurgy, welding and compression technology enabled the building of long-distance pipelines, such as the 1,600 kilometre pipeline from the Texas Panhandle to Chicago. From 1940 to 1970 natural gas went from around 10 per cent to over 30 per cent of total US energy consumption. Despite significant growth in nuclear and renewable energy consumption since the 1970s, in 2010 natural gas still accounted for one quarter of all energy consumed in the US. Gas from the furthest corners of the continent could now be brought to homes in the urban centres of US, providing a cheap source of energy for heating and cooking. It could also be brought to electricity generating stations. Today, that is the principal use of natural gas; it is an efficient way of making the electricity we all need.[96]

Pipeline politics

Where pipelines start and finish can be restricted by both geography and politics. The US, however, was ideally suited in both respects for the emerging gas industry. Pipelines could easily be built over large, flat expanses of land and could be linked to large consumer markets by crossing domestic state boundaries rather than national boundaries. But America is the exception rather than the rule.

'How do I know that when the pipeline is built, China won't turn around and cut the price they pay for gas to only half of our original agreement?' said Vladimir Putin to me in 2004, when I suggested to him the possibility of constructing a gas pipeline from the super-giant Kovykta gas field in the east of Siberia into China. Far from any other possible gas market, China, with ever growing demand, seemed the obvious choice. Surprisingly, there were few pipelines crossing their shared border, despite the fact that Russia is the world's biggest hydrocarbon producer and China is the world's largest energy consumer.

'But how do I know that when the pipeline is built Russia won't start charging us twice what we agreed for gas?' was Premier Wen Jiabao's response when I approached him in search of a solution. I had explained Putin's concerns, and hoped we could find some way to guarantee that China would not backtrack on the agreement. Building a pipeline is very costly. To be financed and to get a return on the investment, a pipeline needs a guaranteed long-term contract between the producer and buyer of the gas. Each party must trust that the other will hold to their part of a deal. For China and Russia, this was not possible.

Through overcoming political and geographical barriers, the use of natural gas has spread across much of the world. And with the ability to trade natural gas internationally, the wealth of nations has grown. But the end of a pipeline is still the end of a pipeline. The inherent restrictions on where gas can be transported from and to often result in a regional, rather than a global, market. The development of LNG has made the market more global, but for an LNG train to be built requires markets to commit to taking the gas for many years. There are, however, surprises as exploration and production technology uncovers new sources of gas. This is the story of the US, which needed to import LNG a few years ago but now has the potential to produce all the gas it needs. That comes about from the successful production of gas from shale deposits, previously thought too technically challenging to tap.

A fracking revolution

This shale gas revolution has largely been brought about by the pioneering work of Texas businessman and philanthropist George Mitchell. In an area known as the Barnett Shale in Texas, Mitchell spent the later decades of the twentieth century experimenting to find an economic way to develop the gas dispersed through the region's impermeable shale rock strata. Natural gas was long known to exist there, but considered worthless as there was no way to extract it. After an initial period of excitement in the 1980s, most other oil and gas companies gave up, but Mitchell continued to have faith in shale gas.

The risk paid off when Mitchell fine-tuned hydraulic fracturing, or 'fracking', technologies to release the gas. The techniques he employed, injecting high-pressure water with sand and small amounts of chemicals to break apart rock formations, were not new and had been used across the oil and gas industry for decades. Indeed, Colonel Roberts, using a different technique, was trying to do the same thing in the nineteenth century. But no one had ever succeeded in using the technology to produce economic quantities of gas. When combined with horizontal drilling, which allows more of a shale formation to be fracked, Mitchell's work finished what Colonel Roberts began.

The US has benefited enormously from the widespread use of these techniques across its many shale basins. At the beginning of the millennium it seemed as if the US would need to import a great deal of LNG. Then, shale gas contributed just 1 per cent of the supply of natural gas. By 2012 it was contributing 36 per cent and reserves of natural gas had almost doubled. At current production rates, the US now has over a century's supply of gas, half in shale and similar unconventional hydrocarbon-bearing formations. Mitchell's innovation had made gas plentiful and inexpensive. It is competing with coal for the production of electricity. And most in the industry cannot remember the last time it was so cheap when compared to oil on an equivalent energy-content basis. Compared with oil, gas has lost around 80 per cent of its value in a decade.

Twelve years after the beginning of the millennium, the US is planning to become a LNG exporter. That has not helped Trinidad. The US market no longer needs Trinidad's gas.[97] Former Prime Minister Manning was right: make sure you use your gas to add as much value in your own country before you export any of it. I suspect that thought will go through the heads of many policy makers.

The successful development of shale gas in the US was enabled by liberal pipeline regulations, subsidies and an abundance of drilling rigs and other infrastructure already in place. These conditions do not exist in the same combination elsewhere in the world, but the development of shale gas and oil is nonetheless beginning to happen, albeit at a slower pace. Countries are dreaming of a new abundance of energy below their feet and some hope that they will become a little more energy self-sufficient.

However, the growth of shale gas development in the US and elsewhere faces great challenges from surrounding communities, environmental NGOs and other pressure groups. They are concerned that fracking can cause earth tremors; that too much scarce water is used and that after it is used it is contaminated and disposed of incorrectly; that the chemicals used in the fracking process contaminate natural aquifers; and that the natural gas released from fracking can get into their drinking water. Dramatic images of US citizens setting fire to water pouring from their kitchen taps have provoked a lot of concern.

Many of these are fears born out of suspicion and misinformation. The incidences of water contamination investigated in the 2010 movie *Gasland*, for example, have largely been shown by the State of Colorado Oil and Gas Conservation Commission to result from activity unrelated to fracking.[98] Some concerns, however, have a basis in fact. Just as with nuclear energy, this newly applied technology will be feared and resisted unless all stakeholders' concerns are balanced with those of commerce. Ideally, the discussion needs to be based on evidence. Some of the evidence in the US is of bad practices by some operators who have cut corners to improve their profits. Standards and regulations are needed everywhere to avoid the reputation of the industry being set by the weakest players.

Big global changes

Shale and fracking represent the start of a big change in the US energy supplies. What Mitchell had started for gas is now applied to oil (and similar liquids) which can be made to flow from naturally almost impermeable shale beds or other geological formations, so-called 'tight oil' or 'shale oil' fields.

The balance of opinion is that the US could at last become effectively self-sufficient in energy; it would not need to import energy from anywhere except its neighbour, Canada, unless it decided otherwise. Every US President since Richard Nixon has aspired towards that independence, seeking freedom from reliance on apparently unreliable supplies from the Middle East and South America.[99] If this comes about, the political implications will be far-reaching. It means that less oil and gas will flow from the East to the US and the West, and much more will flow from the Middle East into Asia. As my friend and Pulitzer prize-winner Daniel Yergin writes, this is 'nothing less than a rebalancing of world oil'.[100] The US would have many choices as to how it considered its historic energy and defence relationships both inside its hemisphere and in the Middle East. China, too, will need to consider its relationships as it continues to import energy and in particular oil from the Middle East. Domestically, the US has, from shale oil and gas, a cheap and reliable energy supply, which gives it a considerable economic advantage. In short, we cannot know how the US will react to being able to meet its energy needs from its own continent. The choices that it and increasingly energy-dependent Asian economies make will, however, shape the global politics of the twenty-first century.

One important side effect of increasing gas consumption of the US is a drop in its carbon dioxide emissions. Natural gas is 'carbon-light'; for an equivalent amount of power generated, it only produces half the carbon dioxide of coal. As a result, the US's carbon dioxide emissions have fallen by almost half a billion tonnes, or 7.7 per cent, in the past five years, more than in any other country. The growing use of natural gas around the world will be important in reducing the risk of climate change.

Carbon fears and climate change

'The power of population is so superior to the power of the earth to produce subsistence for man, that premature death must in some shape or other visit the human race ... gigantic inevitable famine stalks in the rear, and with one mighty blow levels the population with the food of the world.'[101] Writing in 1798, Thomas Malthus darkly predicted a future of famine, disease and war. He saw these disasters as necessary and natural checks on exponential population growth, which outpaced the steady linear growth of our food supply. These checks, he believed, confined humanity to a life of basic subsistence – the so-called 'Malthusian Trap'. Malthus had, of course, failed to foresee the Industrial Revolution.

Malthus was one of the earliest in a long line of distinguished prognosticators who predicted doom but turned out to be wrong because they failed to appreciate the capacity of technological innovation to change the world.

In 1972, the Club of Rome, an international think-tank, famously published a report entitled *The Limits of Growth*.[102] Assuming no technological or political change, it looked at the dangers of rapid population expansion in a world with finite resources, forecasting, among other things, severe shortages of food by the turn of the millennium. This modern-day Malthus failed to anticipate the consequences of the Green Revolution, which has led to a doubling of wheat production over the last forty years.[103] Today, food consumption per head is around 20 per cent higher than in 1972.

Marion King Hubbert and other peak oil pessimists, including the Club of Rome, have also so far been proved wrong. When forecast demand rises above known supply, prices rise, and technology is unleashed to tap new sources of oil or to put existing sources to better use. In 1980, proven oil reserves were around 650 billion barrels. Since then, we have produced over 700 billion barrels, and now have at least 1.5 trillion more in reserve.

Today, the world's pessimists and doom-mongers have settled on a new fear: climate change. They predict a Malthusian catastrophe of famine, disease and war on a global scale. Is there any reason to believe they are right this time, when pessimists have been wrong so many times before?

On initial inspection, there are two reasons to give this new generation of pessimists some credence: a compelling scientific case for anthropogenic climate change, and humanity's apparent inability to address the problem. There exists a strong scientific consensus that human activity is causing the earth's temperature to rise, with extremely severe consequences for human beings and the environment.[104] Energy use and greenhouse gas emissions continue to grow rapidly, with no prospect of a global agreement to control them. As a species we appear incapable of cooperating on tackling this existential challenge.

Science and the risk of climate change

I first began to be aware of the actual risk of climate change in 1992 at the Earth Summit in Rio de Janeiro.[105] More than a hundred heads of state were gathered there to discuss, among other things, what to do about global warming. The science was not new, but this was the first time it had really hit the global agenda.

The discussions at Rio were, on the face of it, entirely at odds with the success of the oil industry and BP. But it was obvious that climate change was not an issue that could be ignored or wished away: I wanted to understand much more and to work out whether BP could 'square the circle'. It had to do something significant towards lowering the risk of climate change.

After Rio, I held meetings with a wide range of scientists and experts in a bid to get to grips with the issue. In the end, one man did more than any other to convince me. John Houghton was chairman of the Intergovernmental Panel on Climate Change (IPCC) scientific assessment group.[106] He spoke in the language of probabilities, a language I had been trained in and understood, and persuaded me that we could no longer be passive about spiralling global emissions.

Climate change is often presented as a simple, linear process of cause and effect: increasing energy consumption will increase the carbon dioxide concentration in the atmosphere, which will cause the earth's surface temperature to rise, which will have catastrophic consequences for our environment. But, as I came to understand it in my discussions with John Houghton, there are many possible outcomes at each step in the chain of causation, and each is subject to significant margins of error.

Modelling a system as complex as our atmosphere and climate, based on assumptions about even more complex systems of human activity, leads to a great degree of uncertainty. The ultimate effects of climate change on the environment, human life and the economy are even harder to estimate. The IPCC calls it 'a cascade of uncertainty'; we do not know exactly how the climate will change or what the consequences of that will be.[107]

But that does not justify inaction: uncertainty does not imply ignorance. It is difficult to know what the world will be like in a hundred years' time, but we can make estimates, acknowledge the inherent uncertainties and act appropriately. And work continues to improve our understanding and reduce uncertainty. Karl Popper once described all science as being provisional.[108] It is a central tenet of the scientific method that all findings can, and should, be questioned. Our knowledge about the connections between human activity, the atmosphere and the environment must be subject to constant challenge and criticism.

The scientific case for anthropogenic climate change is now overwhelmingly powerful. Even fifteen years ago, it was an idea that needed to be considered seriously. That is why, on 19 May 1997, I gave a speech to announce that BP would take action against climate change.[109] Standing in the blistering sun in the Frost Amphitheater of Stanford University, California, I explained that the time had come for BP to 'go beyond analysis to seek solutions and to take action'.[110] On that day, we broke ranks with the entire oil industry.

Over the next decade, the issue of climate change continued to gain momentum. Several leading politicians worked to put climate change on the agenda. US President Clinton and Vice-President Al Gore pushed for the passage of the Kyoto Protocol, the first and only treaty to set binding emissions targets on participating countries. The protocol was a great beginning but suffered some fatal flaws. It did not apply to developing nations, including China and India, and was never ratified by the US or Australia. Gore went on to win the Nobel Peace Prize for 'efforts to build up and disseminate greater knowledge about man-made climate change', notably through his documentary *An Inconvenient Truth*.[111]

In California, the state governor Arnold Schwarzenegger led one of the most ambitious climate-change agendas in the world. In 2006, I found myself sitting in Schwarzenegger's famous cigar tent in the Sacramento Capitol's Quadrangle. He wanted my opinion on a forthcoming bill to reduce greenhouse gases in the state by 25 per cent from 1990 levels by 2020 and 80 per cent by 2050. While the exact targets were ambitious, Schwarzenegger demonstrated that climate change warranted government action rather than just discussion that might be had on a talk show.

In the UK, Tony Blair chose climate change as a focus for the 2005 G8 summit at Gleneagles, Scotland. This, I believe, was the high-water mark for political interest in climate change at the international level. Blair convinced the leaders of the world's eight major economies of the serious risks that climate change posed to our common future, but no agreement was reached on practical actions to take. Despite public concern and real determination among the world's leaders, no global action was taken. The rhetoric had got ahead of reality.

International failure

Since then, the prospects for a meaningful international agreement on climate changed have diminished with each passing year, and with each summit of the UN Climate Change Conference. The 17th Climate Change Conference was held in December 2011 in Durban, South Africa. Although promises were made to work towards a binding agreement in the coming years, nothing substantial was achieved.[112] Heads of state were more concerned with the intensifying crisis in the eurozone. Once again, the issue of climate change was kicked into the long grass.

So why has humanity, faced with the greatest potential challenge to its existence, failed to agree on a solution?

I believe that the problem is one of human behaviour when facing a danger that is uncertain and far off. Anthony Giddens, a distinguished British social scientist, writes that we do not take action against climate change because the dangers posed are not apparent in everyday life.[113]

Electorates do not cope well with probabilities, forecasts and sacrifices; they choose to believe those who tell them that everything will be all right, whatever the reality.

As governments push on with austerity programmes, and as slow growth and inflation erode the standard of living in much of the world, many voters see action on climate change as an unaffordable luxury.[114] The effects of climate change are uncertain, far off and dispersed, unlike the bills they have to pay today. Democratic politicians, elected for short periods of office, naturally reflect those sentiments. Climate change is something to deal with once the economy is sorted out, or once they are re-elected.

But the problem is deeper than mere short-termism. Even if we could overcome our irrationality and act to protect our long-term interests, a global agreement on climate change would not be straightforward.

The problem is one of collective action: the 'tragedy of the global commons'.[115] Carbon dioxide does not respect national boundaries and so the damage caused by pumping more carbon dioxide into the atmosphere is shared by everyone. If one nation reduces emissions but others do not, it still faces the same risk of climate change. While it is in our global interest to reduce emissions, no individual nation will bear the cost without a guarantee that others will follow suit. It is simply cheaper and easier to be a free rider.[116]

Change therefore requires common agreement on how emissions reductions should be shared. But the costs and benefits for each nation are different. Small island nations fear the consequences of rising sea levels and so demand the toughest emission targets. Expansive nations in the far north, such as Canada and Russia, may actually benefit from a warming climate as their land becomes more fertile and their mineral resources more accessible.

The greatest divergence exists between rich and poor nations. Developing nations, most importantly China and India, argue that climate change is not a problem of their making. China emits more carbon dioxide than any other nation, but its per capita emissions are only a third of those in the US. Access to cheap carbon-based energy was the basis of development for all the world's prosperous economies. Why, China reasonably asks, should they be denied the development that the West had enjoyed?

One Chinese delegate at the Kyoto Summit described the difference in viewpoint as: 'What [developed nations] are doing is luxury emissions. What we are doing is survival emissions.'[117] Developed nations, and most importantly, the US, argue that an agreement which lets the developing nations off the hook might place them at an economic disadvantage to their increasingly powerful global competitors.

Faced with divergent national interests and a problem that is uncertain and far off, we should have known better than to pin our hopes on an internationally binding agreement.

We have been here before in our attempt to control the spread of nuclear weapons; it is only in a nation's interest to give up nuclear weapons if they are sure that all others will too. Even faced with risk of nuclear Armageddon, we have failed to reach an agreement to eliminate nuclear weapons or even to prevent their spread.

Climate change presents a collective action problem of the very greatest scale. It requires every country and almost every individual on earth to change the way they live, on the promise that others will do the same, in order to protect the interests of humanity decades from now. That is a vast challenge, well beyond the capacity of democratic institutions at the international level.

Surprisingly, even in the absence of a global agreement, change is already happening, driven by technological innovation in the energy system, and a chaotic patchwork of political progress, emerging from local, regional, and national interests.

The technology of carbon

At the beginning of the Carboniferous period, 360 million years ago, the first primitive forests absorbed carbon dioxide from the earth's atmosphere. Over time, through death, decay, heat and pressure, these forests were transformed into five trillion tonnes of carbon fossil fuels. Since the industrial era, humanity has released around 5 per cent of that back into the atmosphere as carbon dioxide.

Humanity, driven by its desire for energy, has changed the distribution of carbon far more rapidly than ever before. And this distribution will continue to change unless we change how we use energy; by 2050, there are likely to be two billion more people on earth and global energy consumption will almost double.

Technology offers us three ways to rebalance this: energy conservation; low-carbon energy; and the creation of carbon sinks.

First, we can use less energy. If we are unwilling to reduce our standard of living, a reduction in energy consumption requires increased efficiency. In many circumstances that is relatively straightforward and economically attractive, as efficiency gains often deliver significant savings. Innovations in digital communication, such as video calling and collaborative distance working, may be particularly significant.

However, energy efficiency is not the silver bullet that it may appear to be. When something becomes more efficient people use more of it.[118] And when people save money through energy efficiency, they spend the money on other products that use energy. Energy efficiency is not the same thing as energy conservation. Getting people to conserve needs more than the application of technology; it needs education, incentives and potentially regulation.

The second technological tool to restore carbon balance is to reduce the carbon intensity of energy production. Most exciting is the development of zero carbon sources of energy, including renewables such as wind and solar power.

Only a decade ago, these were developmental niche technologies, prohibitively expensive and deployed only on a tiny scale. But in 2010, renewables accounted for almost half of the new electricity capacity installed across the globe. Over the past seven years, there has been a fourfold increase in wind power capacity, while solar power capacity has increased almost thirteen times over. That rapid growth has delivered vast increases in the scale of production and prodigious learning that has improved efficiency and driven down costs. In 2008, for example, solar photovoltaic cells cost $4 per watt of capacity. Today the same capacity can be manufactured for less than $1. Away from the international conferences in Copenhagen and Durban, there has been a quiet revolution that is just starting to transform our energy system.

Nevertheless, renewables are still a small proportion of the energy mix and carbon fuels will be with us for some decades yet. Reducing carbon intensity therefore demands changing the mix of fossil fuels, away from coal and oil towards natural gas. It produces, for an equivalent amount of electricity generated, half the carbon dioxide of coal. Coal has met almost half of new energy demand over the last decade, but the recent boom in cheap natural gas production could soon cause gas-fired electricity generation to replace coal-fired power stations. Gas power stations also have the advantage of being able to ramp up their output quickly, to meet falls in renewable supply when the sun is not shining or the wind is not blowing.

Replacing oil is more difficult, as its energy density makes it uniquely useful in cars and aeroplanes. Natural gas vehicles have come a long way since the 'da qi bao' ('big bag of gas') bus used in China in the 1960s. Giant gas-filled grey bladders would sit precariously on the roof, needing constant refilling as they quickly deflated and sagged over the side of the bus. Natural gas will continue to be used for transportation and if a unit of energy from gas sells for a fraction of that of oil, as is now the case in the US, it will become more ubiquitously used. Electric- and hydrogen-powered vehicles are also possible solutions, but a costly investment in infrastructure is required before their use can become widespread.

A more immediate alternative to replace diesel and petrol is biofuel. Instead of relying on processes which take many millions of years to convert plant and animal remains into crude oil which we then refine into diesel and petrol, we can use a different and faster method: growing crops and refining them into fuel. Carbon dioxide is absorbed in their growth and emitted in combustion. In the US, laws passed in a drive to reduce carbon dioxide emissions mean that most petrol at the forecourt contains 10 per cent ethanol made from maize. There are dangers here: crops for biofuels must not reduce the global supply of much-needed food.

The third and final way to reduce levels of carbon dioxide in the atmosphere is to trap it in carbon stores. Most obviously that means preserving, restoring or expanding existing carbon stores in forests and particular soils. But there is a chance that we may also be able, in time, to create artificial stores of carbon, by intervening in the combustion of fossil fuels, capturing carbon dioxide before it escapes and burying it underground.

In November 2010, I visited the Huaneng Gaobeidian Power Plant near

Beijing to see China's pioneering carbon capture technology in action. The project is the first plant in China to demonstrate the carbon capture process.

A complex array of steel pipes takes up a space about half the size of a football pitch. Inside the pipes, a fluid freezes carbon dioxide escaping from the coal furnace below. As I watched, two workers clad in boiler suits stepped forward and, opening a hatch on the side of the apparatus, pulled out a container of frozen carbon dioxide. Holding some in my gloved hands, I watched it quickly melt and disappear into the atmosphere.

However, the Huaneng project is still just a demonstration, capturing only 3,000 tonnes of carbon dioxide a year, less than a millionth of China's total emissions. It will be a long time before carbon can be captured on a significant scale.

Moreover, capturing carbon is only half the story: somewhere has to be found to store it.[119] The possibilities are varied and imaginative, ranging from storage in rock formations to the creation of large carbon dioxide lakes three kilometres under the sea, the point at which carbon dioxide sinks in seawater. Many people feel uneasy about the idea of storing vast quantities of carbon dioxide inside the earth, fearing what could happen if a carbon store were accidentally released. The same worries over waste storage that have slowed the growth of nuclear power are now coming to bear on fossil fuels.

Carbon capture and storage technology will only have an impact on carbon dioxide emissions if it can be made economically competitive. Energy that would otherwise be used by consumers and industry has to be used to capture carbon. Huaneng sell some frozen carbon dioxide for use in fizzy drinks and the dry ice used at rock concerts, yet the market for those products is very small compared to total carbon dioxide emissions. A more plausible alternative is to inject carbon dioxide into oil reservoirs to recover the last remaining amount of oil.[120]

Bottom-up politics

So human ingenuity presents a wide range of technical solutions: products that reduce our need for energy, new sources of energy that emit less carbon, and methods for trapping carbon that would otherwise be released. We have the tools at our disposal to control carbon's destruction of the planet.

But more than that, we have the social structures to make it happen. The low-carbon revolution is already taking place, not through grand multilateral agreements, centralised decisions and vast new financial mechanisms, but, rather, through a messy, chaotic, stuttering, bottom-up collection of small changes.

At every level – country, region, company and household – changes are being made where they align with the interests of those involved. Few are dramatic, but in aggregate they are significant. Patchworks of nations with common interests can work together to set emissions targets, introduce a price for carbon and support low-carbon technologies, without giving any member a competitive advantage. Europe is perhaps the best example of where this is happening, with common agreement on targets for the amount of renewable energy generation and a scheme for trading emissions.[121]

Individual nations, often driven by concerns about energy security and local pollution of air and water, are also taking unilateral action. More than a hundred countries around the world, both developed and developing, now have support for renewable energy. China, already the world's largest consumer of clean energy, aims to deliver 15 per cent of its energy from renewables by 2020.

Wherever governments are able to explain the benefits that renewable energy can bring, and to convince the people that, as costs fall, subsidies can be phased out, renewables will continue to grow. Where governments can offer financial support that is clear and credible, investment will follow. Over the last seven years, investment in clean energy has grown every single year, at an average rate of 30 per cent.

At the level of the household, Western consumers are changing their habits: consuming less energy, both to save money and protect the environment. Concern about climate change feeds through to spending decisions, and puts pressure on businesses to improve their environmental performance.

Across the world, small coalitions are forming in which change is possible. It will not be as comprehensive, as efficient or as effective as a global agreement, but crucially it is politically feasible. Change is happening, and that should give us hope.

That, however, is not enough. We must find a group of leaders who have a common view of purpose, namely to limit, as much as possible, the risk that climate change will have dire consequences for humankind. They need to encourage all the activity that is happening at country and grassroots levels so as to take us to the point where the possible temperature rise is at least moderated. Since the effectiveness of leadership is context-dependent, they need to pick the right economic and political conditions to go further towards their shared purpose. And, failing that, they need to prepare us to adapt to a different set of climatic conditions. Carbon has served humankind very well; we must ensure it does not exact an impossible price for that.

GOLD

El Dorado

The highlight of my first visit to the Museo del Oro in Bogotá, Colombia, in the late 1980s was the magnificent gold raft of El Dorado, the Gilded King.[1] Sitting in the centre of the raft, the King wears an elaborate head-dress and is surrounded by twelve smaller figures. Some of the simple human forms sit on the edge as if to row the raft, while others, decked in ornate jewellery and wearing masks that resemble jaguar heads, attend to the King. The raft represents the ritual thought to be performed at Lake Guatavita, to the north-east of Bogotá, in which the new leader of the Muisca tribe was received.

As the dawn sun rose, the King would be stripped naked and his entire body painted in fine gold dust. Reflecting the morning light, the King boarded the raft while on the banks of the lake people of his tribe danced, sang and played drums. At his feet the principal chiefs placed more gold, jewels and emeralds before rowing the raft towards the centre of the lake. As they reached the middle, and the shouts from the edge of the lake reached fever pitch, the raft stopped and a flag was raised for silence. The Gilded King, who had been sitting perfectly still and quietly throughout, then made his offering to the lake, casting his riches into the water. As the last gold piece disappeared, the flag fell and the festivities continued in cele-bration of the new ruler. The naked gold man continued to gleam in the sun and so connected to the source of all life.[2]

El Dorado simply means 'the gilded man', yet since the name of the Gilded King was first described by the Spanish historian Fernández de Oviedo in 1541, it has been embellished, exaggerated and has taken hold of

imaginations on every corner of this Earth. El Dorado grew from a gilded man into a mythical city and eventually an entire empire made of gold; it became a dream and a wild aspiration that was always to be found over the next mountain ridge or across the next ravine.

The raft was found in 1969 by three local farmers in a small cave to the south of Bogotá, the capital of Colombia. Alongside caves, the Muisca worshipped mountains, lakes and lagoons. Temples were built at these sacred sites and some tribes even believed that man had emerged from the depths of a lake. But the Muisca's most revered deity was Xue, the Sun god, who represented civilisation and wisdom.[3] Bringing warmth, light and the actual days in which they lived, their faith in the Sun was unswerving. They worshipped the Sun and therefore gold, which is a reflection of the Sun, in both its colour and physical endurance. Gold, reluctant to react with other atoms in the air, does not tarnish; its brilliant glow is as constant as the rising of the Sun each morning.[4]

When I was running BP's exploration business, I often visited Bogotá. It is a traditional Spanish colonial city, centred on a wide plaza that is bordered by a beautiful Presidential Palace. After several years of effort, BP discovered the super-giant Cusiana oilfield in the eastern province of Casanare, a remote and dangerous area 200 kilometres from Bogotá. On one of my visits, Richard Campbell, BP's manager, gave me a piece of pre-Columbian terracotta sculpture. It was a small seated figurine from the ancient Quimbaya civilisation of Colombia who lived on the mountainous slopes of the Cauca River between the first and tenth centuries AD. I was intrigued by the object: whom did it depict? And what religious significance might it have had for the Quimbaya? I got hold of book after book and, as my interest grew, I began to add to my small collection of pre-Columbian works. It was then that I decided to visit the Museo del Oro. Walking into the museum for the first time, I was captivated by the glimmering brilliance of the gold objects. Gold has a unique, sublime quality that marks it out from even the most ornate and accomplished works of stone and terracotta. There is something calming, even comforting, about being bathed in the glow of gold. For the first time in my life I began to appreciate this element's special attraction beyond a store of wealth. Gold holds universal allure. Starkly contrasting cultures have always placed great value on gold.[5] But when cultures collide, the competition created by a desire to own it can have devastating consequences.

In 1537, the Spanish conquistadors first encountered the Muisca people of the Americas. This sophisticated agricultural community supported more than a million inhabitants centred on Bogotá and expanding over an area of around 25,000 square kilometres in the high plains of the Eastern Cordillera Mountains. Swiftly, and with little mercy, the conquistadors took control of these territories. Mounted on horseback with swords and shields, the Spanish easily outclassed the Muisca war-riors who were armed only with weapons made from fire-hardened wood. Since Columbus's arrival on the small islands of the Bahamas in 1491, a similar fate had befallen each native community encountered by the con-quistadors. Whether facing small Caribbean island tribes or the expansive empires of the Aztecs and Incas, the Spanish military force was unbeat-able. The conquistadors had followed Columbus in search of fertile land and a trade passage to the South Sea, but most of all they sought gold. As King Ferdinand of Spain commanded: 'Get gold, humanely if possible, but at all hazards – get gold!'[6]

No story illustrates the conquistador's insatiable lust for gold more than the capture and murder of the Inca Emperor Atahualpa. In November 1532, Francisco Pizarro stormed the Inca provincial town of Cajamarca. As the Emperor was brought forward on a golden throne, Pizarro gave the signal for his men, who had been hiding in houses around the town square, to ride out with firearms blazing. The Spanish were greatly outnumbered by the Inca warriors, but they killed them 'like ants' as they ran from the stampeding horses and gunpowder explosions.[7] Atahualpa was taken prisoner and, fearing for his life, he sent a message out across his empire: the Spanish must be granted both freedom of passage and as much gold and jewels as they desired. He promised Pizarro that he would give them enough gold to fill a room seven metres long by five metres wide. Every day more Incas would arrive carrying gold pitchers and jars, each filled to the brim with pieces of gold. Despite doing everything he could to appease his captors, Atahualpa was tried for treason and sentenced to be burnt at the stake. At the last minute he agreed to convert to Christianity and was spared the fire; instead, he was simply garrotted. Pizarro then swiftly moved through the empire to capture Cuzco, the Inca heartland.

The gold wealth of the Muisca and Inca people was beyond the wild-est expectations of the conquistadors. It provided a foundation for the increasingly extravagant vision of the gold riches of El Dorado. In 1541, the first expedition set out in search of El Dorado from Quito, in modern times the capital city of Ecuador, led by the city's charismatic Governor Gonzalo Pizarro, brother of Francisco. Gonzalo rounded up over two hundred gold-hungry Spaniards and twenty times as many native slaves. On horseback, they headed east out of Quito into what, as he would soon learn, was one of the wildest and most hostile regions of South America. Cutting through the dark, dank and impenetrable forest, Pizarro's men marched blindly onwards for weeks. They tangled with crocodiles, giant snakes and jaguars; with every stroke of the machete, insects would cascade down on them from above. Trudging further and further into the inhospitable marshes, many of Pizarro's men died from diseases never before encountered. With no food, and few native slaves remaining alive, Pizarro was forced to turn back.

The Spanish conquistadors conducted a ruthless campaign, tantamount to genocide. Tens if not hundreds of thousands of natives were murdered as they took control of their territories. Even more natives died from diseases, such as measles, brought from Europe and against which they had no immunity. The native population was close to being exterminated. Legal records of cases brought against conquistadors provide a horrific insight into their widespread crimes. In the town of Cota, the local chief gave Juan de Arévalo less gold than he demanded and so he 'destroyed it, killing many Indians in it, cutting hands or noses off others or breasts off the women and cutting the noses off small children'.[8] When Gonzalo Pizarro arrived in a remote village, he asked about a city made of gold; if the villagers failed to give him the answers he wanted, he had them tortured or impaled with a stake inserted between the legs emerging at their head or simply burnt alive.[9]

Pizarro led the first of what would become many failed expeditions in search of El Dorado.[10] Over the next three centuries, thousands more men, among them Sir Walter Raleigh, would march through the dangerous jungles and grasslands of Colombia in their pursuit of El Dorado. More ground was covered and the legend became increasingly discredited, but enticing rumours continued to pour into the Spanish colonial towns. As late as the early 1900s the Krupp family, the 'iron barons', organised an expedition into the Mato Grosso region of Brazil in search of the Gilded King's legendary wealth. Like all those before, the expedition ended in failure.

Both the Spanish conquistadors and the Incas valued gold, but the Spanish lust for it was wild and insatiable. The Incas could not understand this since, while they endowed gold with religious significance, they had few uses for it. For them, labour was the unit of currency and gold was not a means of payment. Gold is an impractical metal, too soft and heavy to be used in tools and weapons; gold ploughs make no furrows and gold swords take no edge. The Incas saw more value in the iron objects brought to the Americas by the Spanish. For the Spanish more gold simply meant more wealth, and more wealth meant more power. The objects of the religious and artistic heritage of the Americas were melted down and shipped back to Spain, in convoys of sixty ships, each carrying 200 tonnes of gold, simply to be fashioned into coins.[11] That windfall left Spain almost as quickly as it arrived. Much was wasted on extravagant luxuries from the East and on funding equally extravagant wars. Easy money provided little incentive to work, and industry at home stagnated, made worse by emigration of the workforce to the Americas. Spain was quickly led into a spiral of debt and ultimately bankruptcy. As in the late 2000s, easy access to 'free money', today in the form of cheap credit, weakened their economy.[12]

The systematic pillaging and melting down of pre-Columbian gold wiped out a big portion of the cultural legacy of South America. In 1939, the Museo del Oro in Bogotá was founded with the aim of preserving what little remained. The founding object of the museum is a small gourd-shaped object called a *poporos* made by the ancient Quimbaya people. A *poporos* is used to store and crush lime (the mineral, rather than the fruit), which is chewed with coca leaves to increase significantly the release of cocaine-like compounds. It has a particularly sensuous shape, the bulbous curves evoking human sexual organs and fruit. It is a magnificent demonstration of the art of the goldsmith, so much so that I decided that I should find one for my collection. Many false starts and offers of suspiciously found or copied *poporos* later, I acquired what I wanted. The gold work of the Quimbaya influenced much of the subsequent sculpture in Central and Mesoamerica. But these shapes also influenced the great masters of the modern age. The forms were absorbed into the work of Picasso and Matisse, and, in Britain, into the work of Henry Moore and Jacob Epstein. Little reminders of greatly sophisticated shapes of past gold work are everywhere.

Gold casts an extraordinary spell. One of its great attributes is that it is incorruptible, something which has been known since the time of the early pharaohs. The ancient Egyptians revered gold as the unblemished flesh of the Sun god Ra, just as the Muisca, 2,000 years later, adorned their new leaders with gold dust and jewellery to symbolise their divine connections. Only recently, however, in 1995, did two scientists working at a university in Denmark work out why gold does not tarnish.[13] It is because of how the electrons on the surface of gold are distributed. They effectively deny other foreign atoms, such as oxygen or sulphur, the chance to bond. Gold is also a great conductor of electricity. Its power to conduct and not to tarnish makes it a superb way of connecting small electronic components. You can see this on a SIM card in a mobile phone. Electrical applications account for almost 20 per cent of all gold not used for bullion, coins or investment.

The bulk of the rest, some 80 per cent, is fashioned into things to adorn the body. It is so malleable that one gram can be drawn into a wire almost two and a half kilometres in length or be beaten into a one square-metre sheet. There is almost no end to the variations of shapes and decorations that great goldsmiths can make. The Mold gold cape in the British Museum, London, beaten from a single ingot and profusely decorated, 4,000 years ago, is one of several early masterpieces.[14] The Scythian gold jewellery, shining in its case in the Hermitage in St Petersburg, cannot fail to amaze anyone.[15] The gold souks in Iran and the Arabian Gulf sell items both of beauty and value, since a dowry may come simply in gold. I remember the Bakhtiari women in southern Iran both sewing gold coins of their inheritance into the hems of the skirts and wearing large numbers of bracelets with the characteristic yellow colour of pure gold. And on St Mark's Square in Venice there is still a shop, which traces its trade back many generations, making and selling magnificent gold necklaces and rings. Gold jewellery has been an important part of ritual and decoration for millennia. The colour can please everyone. Perhaps that is why it makes women from Hong Kong to Cap d'Antibes believe that they are as alluring as the gold they are wearing. That allure can drive men into a fever, as happened in California.

Gold fever

On the morning of 24 January 1848, James W. Marshall took a routine walk to inspect the sawmill he was building on the southern fork of the American River, California. The mill, Sutter's Mill, was nearing completion, but the water channel was still not wide or deep enough. Each evening Marshall would open the gates of the channel to allow the river to carve out a path while his workmen slept. That morning, however, there was something different about the bedrock uncovered by the water during the night.[16] A glimmering particle caught Marshall's eye. At first he thought it was just a piece of quartz catching the dawn sunshine, but then he spotted several more of these unusual particles, some as big as grains of wheat. 'I picked up one or two pieces,' he recalled, 'and examined them attentively; and having some general knowledge of minerals, I could not call to my mind more than two which in any way resembled this: sulphuret of iron, very bright and brittle; and gold, bright yet malleable. I then ... found that it could be beaten into a different shape.'[17] Back at the living quarters, Marshall's workers were sitting down to breakfast when he burst in: 'Boys, I believe I've found a gold mine.'[18] His find sparked the California gold rush which transformed the lives of thousands of men and women who swarmed to the region in the hope of finding their own El Dorado.

It was impossible to keep Marshall's discovery a secret but it was over a month before anyone else turned up to search for gold at Sutter's Mill. Gold had been discovered in California before, but it had never amounted to much. Residents were sceptical and they believed they were better off putting their efforts into California's thriving agricultural industry. That was until they saw the fruits of the discovery for themselves. One resident recalls a prospector opening a worn sack and out tumbled gold, 'not in dust or scales, but in pieces ranging in size from that of a pea to a hen's egg. I looked on for a moment; frenzy seized my soul; unbidden my legs performed some entirely new movements of Polka steps ... piles of gold rose up before me at every step ... in short, I had a very violent attack of the Gold Fever.'[19]

The fever swept through California and over to the eastern states. When President James Polk reported to Congress in December 1848 that the rumoured gold wealth of California was a reality, the whole world became gripped.[20] Immigrant prospectors travelled from all corners of the world: the East Coast of the US, Chile, Australia, China and Europe. They arrived in California and lived in overcrowded makeshift settlements, racked by disease and lacking the most basic of amenities. But everyone was focused on gold; no one took the time to build permanent shelters or infrastructure. They lived in extraordinary hardship. They inflicted immense cruelty on the native inhabitants, who regarded them as invaders. Whole native villages were surrounded and captured at night, often in retribution for attacks on a single prospector.[21] They did not, however, steal their gold from the natives but worked to get it out of the ground. And while these prospectors were similar in many ways to the conquistadors, nothing could surpass the cruelty and tyranny of those Spanish.

Early arrivals in California quickly stumbled across windfalls of alluvial gold, easily digging lumps out of the river bed with their bare hands or a pocket knife. Seven weeks' work provided one prospector with over a hundred kilograms of gold, worth around US $40,000 (over US $1 million today). In the first few months, as the search spread far beyond Sutter's Mill on the American River, a few thousand prospectors took away with them over US $2 million worth of gold (US $60 million today). For many men, the hope of a lucky gold strike went unfulfilled. Surface deposits were soon picked over and so men had to spend long, back-breaking hours panning through pay dirt or digging side tunnels into the river bed. As the easy gold finds dried up, the gold seekers' optimism was gradually worn down. Many had no choice but to stay, scraping just enough gold from the ground to sustain a life in the camps, known as 'mining for beans'. James Marshall, who sparked the Gold Rush, died a destitute drunk.

In the early 1850s, gold became harder to find and its extraction was left to the growing body of industrial mining companies. They followed the gold back to its source digging deep into the river valleys and flattening entire hillsides using jets of water to break off ore. Gold was no longer found in lumps and rock had to be crushed into powder to extract the ever smaller gold particles it contained. The once ramshackle and lawless settlements began to take on a more organised appearance. Many unsuccessful gold seekers turned to shopkeeping. In 1853, a young Jewish-German immigrant to the US called Levi Strauss moved to San Francisco, the gateway to the goldfields, and set up a store. Two decades later he began the production of the world's first denim jeans, the uniform of the gold miners, the Wild West and, in our times, the world.

When Marshall had struck gold in January 1848, California was a small territory not yet having acquired statehood, with a population, of around 20,000, excluding indigenous people. Within a few years it had a population of hundreds of thousands and by 1880 that was nearly a million. Today it is the most populous state in the US, and one of the world's largest economies. In California, the scale and success with which gold was invested was unprecedented. Its discovery led to a period of extraordinarily fast development at the time of the US industrial revolution. California was connected with the booming East Coast, both through immigration and the development of railways. It was transformed from an agricultural to an industrial state. Gold underpinned that development through investment. And while the opportunities for gold prospectors had dried up, California's industrial revolution was providing new ways to make a fortune in the state's nascent industries. It seemed possible for anyone to get rich quick during this period of unparalleled growth.

Gold coins

In ancient Egypt, gold jewellery was high art, worn by pharaohs to symbolise their power and to assert their proximity to the gods. It was used as gilt for divine statues, temples, obelisks and pyramids. Gold's connection with the gods meant it was also extensively used to honour the dead, most famously in the mask and coffins of Tutankhamun. By these divine associations, gold became valued and, as a consequence, widely acknowledged as a store of wealth. To do just that, the Egyptians cast gold ingots as early as 4000 BC. Unlike crops and livestock, gold is neither bulky nor perishable and so it can be used as a means of payment over longer distances and periods of time. It does, however, have some limitations. The value of gold will depend on its purity and there needs to be a way of ensuring this if gold is to be a trustworthy means of payment.[22]

In the seventh century BC, the thriving merchant city of Sardis (the capital of Lydia, the site of which is in modern Turkey) which had an abundant local supply of gold and silver, came up with one of the first solutions to this problem. As the River Pactolus, which passed through the city, meandered on its way towards the Mediterranean coast, it slowed and deposited gold and silver, combined in a naturally occurring alloy known as electrum.

According to Greek mythology, the river bed was given these riches when King Midas bathed in the waters to remove his cursed golden touch. The proportion of gold and silver in electrum varied considerably and therefore its value also varied. In its raw state, it was not a trustworthy medium for payment. However, the Lydians made the electrum into coins, which signified that each coin would be redeemable for its face value, regardless of the actual value of the gold and silver from which it was cast.[23] The difficult and often inaccurate task of testing the electrum's purity each time it changed hands was no longer necessary. Merchants could place trust in a transaction, without having to trust the individual with whom they were dealing.

In the middle of the sixth century BC, during the rule of King Croesus, the Lydians discovered a method for separating electrum into almost pure gold and pure silver by heating it with salt. Croesus became the first ruler to issue coins of pure gold and silver, stamping them with the symbols of the Lydian Royal House, the opposing figures of a lion and a bull, representing the opposing forces of life and death.[24] These overcame the big drawback of electrum coins which were only accepted locally; the pure gold and silver coins, struck with the royal stamp, became accepted internationally and that was important for Sardis. Situated between the Aegean Sea and the River Euphrates, it was perfectly placed to take advantage of the growing international east to west trade routes. Croesus's coins improved the efficiency and reliability of the growing number of international transactions and his coins soon spread around Asia Minor. Gold was no longer purely a store of wealth for the rulers, but it had been placed into the hands of merchants. As international trade grew, so did the amount of gold and silver changing hands.

As for Lydia, its wealth was frittered away on luxury goods and on Croesus's ever more ambitious wars. In 546 BC Croesus attacked Cyrus the Great's Persian Empire, but this was a step too far.[25] Within a year he was defeated, but his legacy lived on as his coinage was used by the Persians as they pushed westward. Croesus's invention had turned gold and silver coins into standards, the use of which would, in time, spread across the entire world.

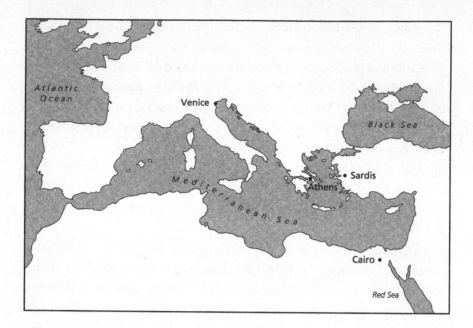

Symbols of civilisation

In my library I have a book of very large engravings made by Giovanni Battista Brustolon based on paintings by Canaletto. One of these images shows crowds thronging on all sides of St Mark's Square, beaten back by men with sticks and dogs. At the centre sits the Doge of Venice, on his *pozzetto*, a type of sedan chair. He has just been inaugurated inside St Mark's Basilica and that forms the backdrop for the scene. And as he is carried through the square, the Doge throws hundreds of gold and silver coins into the crowd, resulting in a scene of chaos. Here, just as in the depths of the South American jungle, men and women scrambled for gold.

That scene, drawn in the eighteenth century, was of a city that, six hundred years earlier, had become the new trading hub of Eurasia. In Venice, as in Sardis, it was gold currency that underpinned the prosperity of the city. In the twelfth and thirteenth centuries, east/west trade increased as the economy of Western Europe grew. Silk, spices and other luxurious items were brought into the trading centres of Genoa, Florence and Venice where, because of their rarity, they fetched high prices. So much gold flooded into Florence that by 1252 there was enough to start mint-ing a new gold coin, the florin. In 1284, during the administration of Doge Giovanni Dandolo, Venice followed and minted its first gold ducat with the same weight and finesse as the florin.[26] The gold ducats soon became the coins that defined value and took that role away from the silver grossi which were also minted in the city. Venice's formidable fleet of ships, used for both trade and war, ensured the spread of the ducat throughout Europe.

The ducat became a European standard, understood and accepted everywhere, and an advertisement of the strength and power of Venice. A decree from the Venetian Senate in 1472 asserted that 'the moneys of our dominions are the sinews, nay even the soul, of this Republic'.[27] The Zecca, the mint where all Venetian coins were cast, was at the city's centre phys-ically and politically. Sitting on the waterfront, the three-storey building stretched along St Mark's Basin with entrances facing towards the doge's palace. Each doge took great pride in the Zecca and its prodigious output of ducats, on which their image was struck. Here for over five hundred years, until the end of the Republic in 1797, the Venetian ducat was minted, retaining the same weight and purity as it had always done.

However, circulating among the fine gold ducats were forgeries and debased coins whose edges had been clipped. Clippers would remove small amounts of metal from the edge of coins, then melt the clippings down and sell the metal back to the mint. Counterfeiters would simply strike imitation coins made of a different, lower valued, alloy. Clipping and forgery began to undermine trust in the Venetian currency.[28] It became so prevalent that the Council of Forty, the supreme court of Venice, declared clipping a sin, abominable to God. Those men caught would be fined, have their right hand amputated and be blinded in both eyes. Women faced life imprisonment.[29]

Venice was not the only place to have this problem. In Elizabethan England, coin clippers could be hanged or burnt alive, but these pun-ishments did little to deter people. By 1695, in the reign of the Stuarts, counterfeit money accounted for around a tenth of all coins in circulation in England. 'It was mere chance whether what was called a shilling was really ten pence, sixpence, or a groat,' wrote Lord Macaulay, the Victorian historian.[30] The following year the Chancellor of the Exchequer appointed Isaac Newton to be the Warden of the Mint. Newton is best known for his falling apples and the laws of motion but from his time at the Mint he left behind a more tangible legacy: modern coins. The position was meant to be a sinecure, but Newton threw himself into the job with immense enthusiasm. As Warden, one of Newton's roles was to enforce the law on crimes committed against the national currency. He became both a detective and a law-enforcement official, applying his scientific genius to the pursuit of London's criminal minds. In an effort to foil the counterfeiters, Newton recommended that all England's currency be replaced. Many older coins in circulation were worn down and thus easy to imitate. However, the Mint had developed a sophisticated new technology for inscribing the edges of gold coins with the words *Decus et tutamen*, meaning an 'ornament and safeguard'. This inscription can still be found on the edges of British pound coins.[31] No seventeenth-century counterfeiters had the equipment to do this and so the Bank of England wanted to replace all the old coins with these newly protected ones. But Newton had an even more important reason for making and circulating new coins. For some time silver had been disappearing from England. Since lower denomination silver coins were needed to pay for everyday transactions, their disappearance was harming domestic business. Newton set about applying his great mind to this troubling economic problem.

A value standard

Britain's coinage system was suffering from a phenomenon called Gresham's law: simply that bad money drives out the good money. The relative value of an ounce of gold to an ounce of silver was fixed by law at a level that made silver less valuable domestically than it was abroad. People could net a profit by melting silver coinage down and then selling it at a higher price on the international market. In this way silver, the 'good' money, was being driven out of the country. Lord Macaulay wrote: 'Great masses were melted down; great masses exported; great masses hoarded; but scarcely one new piece was to be found in the till of a shop.'[32]

Newton understood that mere laws would not be enough to stop the smuggling of silver bullion abroad. He had seen how counterfeiters and clippers risked death for a quick profit. Rather, he recommended the obvious: that the exchange rate between gold and silver be set closer to that prevailing abroad, prohibiting anyone from paying or receiving gold guinea coins at a value other than twenty-one silver shillings. But the adjustment was not quite enough and silver continued to leave the country. Gold coins soon dominated and Great Britain found herself using a *de facto* gold standard, in which the standard monetary unit is defined by a fixed amount of gold. Newton's miscalculation had raised gold to an exalted position; just over a century later, in 1816, the Bank of England introduced the British gold sovereign.

Like the Venetian ducat before it, the gold sovereign was the symbol of the world's most powerful trading nation and became accepted globally as the safest currency. While living in Iran in the 1950s I remember travelling with my father to a remote village to buy a Persian carpet. After bartering with the seller, we drove away with three carpets, which are still on the floor of my house today, for the price of three gold sovereigns and an old suit. Even there in the middle of a distant desert, the value of the British sovereign was known and trusted.

The world's dominant economy had moved on to a gold standard and other nations soon followed. The gold standard was introduced in Germany in 1872 following the Franco-Prussian War, and in Holland, Austro-Hungary, Russia and Scandinavia soon after. The Coinage Act of 1873 put the US on a *de facto* gold standard and France joined in 1878. By the early 1900s only China, parts of Latin America and Persia had not followed suit. These moves were made possible by the sudden growth in global gold reserves from new mines in America, South Africa and Australia. In the first six years of the Californian gold rush prospectors produced nearly US $350 million in gold (US $10 billion today). The discovery of this gold was the impetus to search for new mines across the world. Only sixty years after Marshall's original find, world gold production had increased by one hundred times. Much as the rulers of Lydia and Venice had used a sudden influx of the precious metal to reform their currency system, the great new flows of gold across the globe were used to transform the international monetary system. That system brought many benefits to the nations who were becoming more and more reliant on trade among themselves. This demanded stable currency exchange rates which were achieved by pegging each nation's currency to a fixed amount of gold. Governments guaranteed that anyone could convert their currency back to gold and so the movement of different currencies between different countries was made possible. The international gold standard provided the security and stability needed for the globalisation of capital and trade. It also ascribed a price to gold that was fixed and immutable.

From 1880 to the start of the First World War, the gold standard flourished. In a period of relative global peace and prosperity, the world's reverence for gold supported global economic growth. Just as the gold coins of King Croesus and the Venetian ducat were central to the rise of two great Eurasian civilisations, so the gold standard enabled the economy of the entire world to thrive. But this was not without controversy. The US had used both gold and silver as legal tender and that created all the problems that were evident in the UK. After its civil war it went on to a mono-metallic gold standard. This constrained inflation and facilitated international trade but reduced the money supply. Advocates such as the 'Gold Democrats' and the mainstream Republicans talked of monetary discipline breeding prosperity while its opponents – the mainstream Democrats and the 'Silver Republicans' – wanted the return of silver which would inflate the money, make 'the struggling masses' feel more prosperous at the expense of 'the idle holders of idle capital',[33] and benefit their local silver mining industry. All of this culminated in the great speech in 1896 of William Jennings Bryan who, arguing against monetary discipline, declaimed, 'you shall not press down on the brow of labour this crown of thorns, you shall not crucify mankind upon a cross of gold'.[34] But this great peroration was to no effect. More gold was discovered around the world and the money supply eased.

During the First World War, the gold standard was put under increasing pressure. Huge military expenditure unbalanced budgets, resulting in high inflation, with the result that many governments could no longer guarantee the conversion of currency back into gold. In both world wars, many nations took themselves off the gold standard. After the wars, large public spending for reconstruction placed similar pressures on government finances. Many governments did not want to be restrained by the gold standard, instead preferring a monetary system that allowed them to pursue domestic political aims with freedom. British economist John Maynard Keynes was clear that participating in the gold standard limited a democracy's ability to set its own domestic policies, such as reducing unemployment.

In spite of all of these difficulties, the desire for a gold standard would not go away. The Bretton Woods Agreement was its most successful revival. In the search for stability and unity after the Second World War, the international gold standard was reintroduced in 1944. By agreement, the value of the dollar was pegged to the value of gold. The US was able to do this since its economy was in good shape, having suffered little war damage. But no currency is as enduring as gold. In the 1960s, the previously thriving US economy began to shrink. Large domestic expenditures and the mounting costs of the Vietnam War led to an increasing imbalance of payments. Inflation and unemployment were on the rise and the US's stock of gold was rapidly shrinking. The US could no longer guarantee convertibility of its currency into gold and the Bretton Woods Agreement became a constraint on the government's ability to strengthen the economy.

President Nixon had two options: to implement high taxes, high interest rates and extreme budget discipline, or to abolish the dollar's tie to the value of gold and allow it to float freely in foreign exchange markets. On Friday 13 August 1971, the President went to Camp David in secrecy with his economic advisers, where he made the decision to abolish the gold standard. In the pandemonium that followed, the price of gold kept going up. By 1980 the gold price reached what was then a record high of $859 an ounce, as investors sought refuge from inflation and placed their faith in gold. And just as the price peaked, a lucky find in the Brazilian jungle once again demonstrated the lengths to which humans would go in pursuit of this alluring element.

Timeless allure

In 1980 near the small town of Três Barras in the south of Brazil, a farmer was working the land he had carved out of the surrounding jungle when he noticed a gleaming gold nugget piercing the mud. He sold the nugget at a nearby town and from there the news quickly spread. Soon 10,000 *garimpeiros*, gold diggers, had appeared on the farm in the hope of finding something themselves. Ramshackle settlements sprang up, much as they did during the Californian gold rush. Workers lived in insanitary and generally inhuman conditions. Crime was commonplace; murder and prostitution were rife. Between 1979 and 1980 gold production in Brazil had increased more than fourfold to 37 tonnes, in large part owing to output from the mine at Três Barras.[35]

In 1986, when the photojournalist Sebastião Salgado arrived at Três Barras, now called Serra Pelada, it was better organised. The site had been split into small sections, each dug by a small team of workers. The owner of a section paid a miner just twenty cents for each 50-kilogram sack of pay dirt he carried. When gold was struck in a section, each worker had the right to pick one of these sacks. 'Inside they may find fortune and freedom,' wrote Salgado. 'Their lives are a delirious sequence of climbs down into the vast hold and climbs out to the edge of the mine, bearing a sack of earth and the hope of gold.'[36] The mine was in rough terrain and that meant that mechanised equipment could not be used. Salgado's images captured the workers, looking like ants, as they scrambled up ladders against the treacherous mine slopes. His black and white images appear to be from a bygone age, more akin to the Spanish conquistadors' mines of the sixteenth century than of anything from today. Gold offers the potential to support the development of an economy, but all too often its bounty falls into the hands of a powerful few. At Serra Pelada, a few rich landowners took away the great majority of the mine's output, leaving thousands of workers destitute when the mine closed in 1992.[37]

Human use of gold has not changed greatly since the time of the pharaohs, or of the Colombian Muisca, or of Renaissance Venice. In all these different eras gold was both an adornment and a store of wealth, embodying both power and security. In times of prosperity, wealth was on display and coins changed hands; in times of insecurity gold was hoarded. During the growth of the international gold standard in the late nineteenth century, much of the newly mined gold was placed in underground vaults as bullion. In 1930, Keynes reflected on the disappearance of gold: '[Gold] no longer passes from hand to hand, and the touch of metal has been taken away from men's greedy palms. The little household gods, who dwell in purses and stockings and tin boxes, have been swallowed by a single golden image in each country, which lives underground and is not seen. Gold is out of sight, gone back again into the soil. But when gods are no longer seen in yellow panoply walking the earth, we begin to rationalize them; and it is not long before there is nothing left.'[38] Keynes predicted that gold would lose its irrational allure. But recent history says something different. In August 2011, in the aftermath of the financial crisis that began in 2008, gold reached a record nominal high of near $1,900 an ounce, more than double its price only four years previously.

Miners have always gone to extreme lengths to get gold. I first visited the gold and copper mining pit at Bingham Canyon, in Utah, in 1986, when it was owned by BP. Bingham Canyon is 4.4 kilometres across and 1.2 kilometres deep; it is the biggest man-made hole on the face of the Earth, mined with equipment that made me feel like an ant in a land of giants. Copper was, at the time, of so little value that it was the gold that kept this mine going. The Tau Tona gold mine, four kilometres under the surface in South Africa, is the world's deepest mining operation. Here the pay dirt contains only tiny traces of gold. Each year 500 tonnes of gold are mined in South Africa, requiring 70 million tonnes of earth to be raised and milled in the process. We seem to dig gold out of the ground at great cost, only to put it back under the ground as bullion.

Gold's allure is timeless and we continue to treat it with an almost religious reverence. No currency is tied to it and no trade is facilitated by it but still we want more and more of it. It is still *the* global symbol of wealth and success. Elegant, sun-tanned women adorn themselves with gold jewellery in a contemporary imitation of El Dorado; Olympic champions and Nobel prize-winners still cherish the allure of gold medals; we still exchange gold rings when getting married. Our faith in gold is enduring; our relationship with it has never grown up.[39] Gold does not go away, because it is incorruptible. So where is all the gold of the pharaohs and conquistadors? It was simply melted down and transformed from one form to another. The wedding ring on your finger could be made from the gold from the output of Bingham Canyon or the Tau Tona gold mines. But there is just a chance that in it there is a little piece of the gold riches of El Dorado.

SILVER

S ilver first appeared in my life through photography. I was a child in Singapore and quite lonely because both my parents worked and left me in the care of a nanny. The fact that they both worked in the 1950s was unusual; the company of a nanny was not. To keep me amused I was given a camera and that began a lifelong affair with photography, the basis upon which silver changed my world. It gave me a source of pleasure, a way of enquiring and a way of seeing.

Long before the invention of photography, silver was used, like gold, as a store and symbol of wealth. And this is how I start silver's story. In South America rumours of La Sierra de la Plata, the silver mountains, spread among the conquistadors just as the legend of El Dorado had done previously. Those mountains, they heard, lay across the hot and arid lowlands of the Chaco Boreal. Suffering treacherous climates and committing atrocious crimes against the natives, the Spanish conquistadors arrived in the Cordillera Real mountain range, but, unlike the expeditions that set out in search of El Dorado, their search was not in vain. Rising to a peak of almost 5,000 metres above sea level, the conical Cerro de Potosí rose up in front of them. Digging beneath the barren slopes, they found that the mountain's heart was made of silver.[1]

A mountain made of silver

Around the turn of the sixteenth century, decades before the arrival of the Spanish conquistadors in the Inca territories, Emperor Huayna Capac (the father of Atahualpa who was subsequently brutally murdered by the conquistadors) saw the perfectly conical form of Cerro de Potosí and concluded, somehow, that it must be filled with precious stones and metals. He was right. Tens of millions of years before, volcanic material, rich in metallic minerals, was pushed towards the earth's surface and forced into the rock fractures of the nascent Cerro de Potosí. Here it solidified, forming veins of silver that were both densely packed and highly enriched. Hidden beneath the mountain's barren slopes, these veins would lie in wait, undisturbed until the time of the Incas. According to legend, the Emperor's men had begun to dig into the hillside when a booming voice was heard to say: 'The Lord has reserved this for others who will come after you.'[2] They fled the mountain and gave it the name, Potosí, which in Quechan means 'big thunder'.

Those 'others', it seems, were the Spanish conquistadors. No one knows exactly who first opened the veins of Potosí. Stories abound, such as the Indian who, while chasing an errant llama, stumbled across a silver outcrop on the mountain's slopes. By the time the Spanish arrived in the 1540s, it is very likely that Indians were, in a small way, already mining, but large operations only began after Gonzalo Pizarro arrived in the Inca Empire. By 1545, the conquistadors were cutting deep into the mountain's heart of silver. The small settlement of Potosí, situated in the shadow of the mountain and sharing its name, soon became the centre of Peru and one of the world's biggest and richest cities. When the first census was ordered by the Viceroy Francisco de Toledo around twenty-five years later it had grown from a few small huts to a settlement of 120,000 people.

Situated 4,000 metres above sea level, it is a cold, dry and very windy place. Luis Capoche, a sixteenth-century silver miner and refiner, wrote of the violent gales that 'bring with them so much dust and sand that they darken the air'.[3] But in this barren place, set in the Andes, the Spanish created an opulent city in which grand colonial mansions lined the narrow streets and where merchants sold 'felt hats from France … steel implements from Germany … and crystal glass from Venice.'[4] For entertainment, they built gambling houses, dance halls and theatres. Some seven hundred professional gamblers and more than a hundred prostitutes lived in the city. Fights were a regular occurrence between the many diverse groups that flooded into the city from all over the world and duels became a regular and permitted activity. Potosí was the South American centre of colonial debauchery. It played host to the most extravagant of festivals. In 1608, the feast of the Holy Sacrament was celebrated with six days of plays, six nights of masked balls, eight days of bullfights, three days of fiestas and two days of tournaments. One fiesta included a plaza containing a circus 'with as many different kinds of animals as in Noah's Ark, as well as fountains simultaneously spouting wine, water, and the native drink.'[5] El Dorado's city of gold may have been a myth, but from the hidden wealth of the Cerro de Potosí the conquistadors created their own city of silver. The lavish city life reflected the 700 per cent expansion in production between 1573 and 1585, enabled by new hydraulic mining technologies. Those needed a lot of water the year round. Vast dams and a system of thirty-two lakes containing some six million metric tonnes of water were constructed. In 1592, silver production peaked at 200 tonnes per annum before slowly declining through the seventeenth and eighteenth centuries. In two hundred years the mountain had yielded more than half the world's production of silver, and a total of more than 30,000 tonnes.

The Spanish saying 'vale un Potosí', as rich as Potosí, became an expression of the highest praise.[6] Potosí became a global symbol of wealth, leading the Holy Roman Emperor to grant Potosí the title of Imperial City, presenting a shield with an inscription, translated as: 'I am rich Potosí, treasure of the world, king of the mountains, envy of kings.'[7]

But underneath Potosí's glittering display of wealth lay a darker reality, the effective slavery of millions of Indians. Inca society was based around a system of coerced labour so that the Indians had an 'ingrained acquiescence to the notion of tribute.'[8] Effectively, the conquistadors exploited this social structure which enslaved the native workers. It was kept in place for 250 years. Over 10,000 men were taken each year from their homes in the Bolivian and Peruvian highlands.[9] The Spanish considered them 'animals without masters', sometimes making them work up to half a kilometre underground for twenty-three weeks a year without rest, day and night.[10] Whipping and beating was common practice and this often killed the exhausted miners. Cerro de Potosí came to be known as 'the mountain that eats men', a name it retains among miners today who still risk their lives extracting tin from the ancient mine shafts.[11] A llama is still sacrificed every year at the mouth of the mine in the hope of appeasing the devil that lives inside the mountain. But today the city of Potosí, like the mountain, is a shell of its former self. High in the Andean mountains, with a climate unsuitable for growing crops and disconnected from major South American cities, only the silver mines sustained the lavish city life in the sixteenth and seventeenth centuries. When the precious metal was gone, the people, the gambling halls and the grand fiestas followed. Unlike in California after its gold rush, there were few opportunities for the local economy to diversify. The mountain, riddled with tunnels, has long since been ransacked, the surface a slag heap, coated in worthless, discarded rock that gives Cerro de Potosí a reddish hue.

Along with the gold of Atahualpa and of the Muisca, much of Potosí's silver was shipped to Europe by merchants or as tribute to the King of Spain. But as one seventeenth-century observer remarked: 'What is carried to Spain from Peru is not silver, but the blood and sweat of the Indians.'[12]

The silver mines of Bohemia

Silver from Potosí flowed through the trading centres of Genoa, Florence and Venice. Production at Europe's own silver mines was also booming. In the late fifteenth and early sixteenth century, new mines were discovered in Saxony and the Tyrol and new mining technologies were making many old mines profitable once again.[13] In 1527, during this mining boom, Georgius Agricola became the town doctor in the small Bohemian city of Joachimsthal, close to where he was born in Saxony. Joachimsthal, today Jáchymov in the Czech Republic, sits on the slopes of the Ore Mountains, the centre of Europe's mining industry. There he saw how the science and technology of mining and processing ores was transforming the world around him. The town had only been founded eleven years before Agricola's arrival, but within that time had grown from a population of 1,000 to over 14,000, as a result of the prodigious output of the city's silver mines and the 'thaler' coins minted from it.[14] Much of Agricola's time was spent studying and visiting the mines and smelters on the surrounding hillside. Around 1530 he resigned his position as town doctor and spent the next few years travelling. His subsequent studies came to form part of his work on mining, *De re metallica*, whose authority on the subject would not be surpassed until the Age of Enlightenment.

In this work, Agricola not only explains, in meticulous detail, how to find, extract and process valuable ores but also becomes a lobbyist for the miner and his art. To Agricola, the advantages of mining were all too clear, yielding the ingredients for medicines and paints, and metals for coins. 'No mortal man,' writes Agricola, 'ever tilled a field without [metal] implements.'[15] To those who questioned the relevance of mining, who believed that miners have 'the most bitter and miserable lives', Agricola asked: 'How much does the profit from gold or silver mines exceed that from agriculture? ... Those who condemn the mining industry say that it is not in the least stable, and they glorify agriculture beyond measure. But I do not see how they can say this with truth, for the silver-mines ... remain still unexhausted after 400 years.'[16] But, he says, mining is only profitable 'to those who give it care and attention'.[17] The miner must understand a great many arts and sciences if he is to be successful.[18] He must know how to ensure the health of his workmen and how he may 'claim his own rights' through the law. It almost feels as if Agricola is describing the capabilities of today's mining companies.

Agricola recounts the discovery 'by chance and accident' in the twelfth century of a great silver vein by the River Sala near the town of Freiberg, fifty miles from Joachimsthal, under the control of the Margrave Otto of Meissen, more commonly known as Otto 'the Rich'.[19] For much of the thirteenth century, Freiberg was the centre of European silver discoveries, with one mine taking over when the previous one became exhausted. But this stopped in the second half of the fourteenth century, when fewer and fewer new mines were discovered and the silver stopped flowing into Europe.[20]

Miners following a silver vein eventually reached thin black veins of pitchblende, which we know now as uranium ore, and which usually meant the end of that particular silver vein. The pitchblende at Joachimsthal was very rich in uranium and, much later, was used for the building of atomic bombs. Other mines would be abandoned because they contained 'demons of ferocious aspect', wrote Agricola, which must be 'expelled and put to flight by prayer and fasting'.[21] We now interpret these demons as noxious and explosive gases. The art of mining has made great leaps since his time. We no longer use astronomy to judge the direction of veins, nor do we believe in demons and trolls. The four-kilometre-deep Tau Tona gold mining pit in South Africa and the complex chemistry of rare earth metal extraction are beyond Agricola's wildest imaginings. However, his reverence for the wide benefits that mining can bring to society holds true today. Whether these benefits are realised depends on how we choose to extract and use minerals. Agricola understood that the elements are neither good nor bad: 'For good men employ them for good, and to them they are useful. The wicked use them badly, and to them they are harmful.'[22]

The silver from Bohemia transformed European coinage. Otto the Rich and his successors opened dozens of mints in Germany, casting silver coins that flowed into England in exchange for wool and cloth. Silver coins were minted on an unprecedented scale. They were, for many purposes, more practical than gold ones. Their lower value enabled them to be used for smaller transactions among medieval Europe's burgeoning merchant class. Silver coins even passed through the hands of peasants, perhaps for the first time since antiquity.

The Owls of Athens

The Owls of Athens are thick and heavy silver coins stamped with the helmeted head of Athena on one side and her owl, the symbol of wisdom, on the reverse. First minted at the end of the sixth century BC in Athens, they soon spread far outside its walls. They were used as a tool of foreign trade and became a symbol of Athenian power. Other coins came and went; only the weight and form of the Owl remained constant for more than three centuries. And that consistency ensured that Athenian coins were accepted and trusted throughout the Mediterranean. Only the Venetian ducat of medieval Europe, which was minted for more than half a millennium, would rival the longevity and credibility of the Owl. The first Owls were minted just as Athens was starting to go through a period of unprecedented political and economic growth with democracy firmly established.[23] Athens' prosperity depended on its control of the Aegean seas, its islands and coastal cities, and that was made possible by its formidable fleet. It also depended on silver, much of which came from the silver mined at Laurium, 65 kilometres south of the city. The mines helped Athens rise to become the pre-eminent civilisation of the Mediterranean. Xenophon wrote: 'the Divine Bounty has bestowed upon us inexhaustible mines of silver, an advantage which we enjoy above all our neighbouring cities, who never yet could discover one vein of silver ore in all their dominions.'[24] But in 413 BC, Athens began to lose control of its silver when the Spartans captured a nearby town, creating a fortress from which to occupy the region.[25] A general defected to the Spartans and told them how their occupation could cut Athens off from the silver mines of Laurium.[26] The plan worked, the mines were closed and, in an act of desperation, the city melted down the gold objects on the Acropolis, including eight gold statues of Nike, the goddess of victory. In 404 BC, following a prolonged siege, Athens surrendered to the Spartans. Thucydides described the chronic shortage of silver as a principal cause of Athens' defeat. The silver mines had supported the rise of a great civilisation, but their absence hastened its fall.

Nike had been abandoned. From her, new gold coins were made with the same weight and form as the silver coins they replaced. But because of gold's greater value, each coin was worth twelve times its silver counterpart. This high value made the coins of little use for local retail transactions and they were predominantly used to pay for shipments of goods from abroad. It was silver, not gold, that reigned supreme during Athens' cultural, political and economic heyday. The importance of silver was evident in the historically low ratio between the value of gold and that of silver. In Athens, gold was worth twelve times the value of silver, very similar to a ratio of ten decreed by King Croesus in Lydia. For over three millennia it has fluctuated in the range of 9:1 and 16:1; in 2011, gold was worth at least fifty times as much as silver.

The relative values of silver and gold simply float, based on relative supply and demand. Local variations in the distribution of sources of gold and silver lead to local differences in the ratio of their values, but, as international trade develops, these variations are gradually smoothed out. An ounce of silver was always worth much less than an ounce of gold. That was one of its historic attractions because it could be made into smaller 'change', low-value coins for everyday trade. To make this practical and to ensure that tradesmen knew where they stood, state treasuries or monetary authorities would need to fix, as opposed to let float, the ratio of value between gold and silver. That, time and time again, opened up opportunities for unintended arbitrage at the expense of the state. For example, as early as the Middle Ages, when silver was abundant in Europe, it was profitable to ship the bullion across the Mediterranean to northern Africa where it would be swapped for gold. Silver has always been more abundant than gold and, for state treasuries who wanted to ease the money supply, silver was the thing they used. That 'unsound' monetary practice could never be sustainable. Over time, gold became the preferred international standard of value, it became more in demand and its relative price strengthened.[27] Silver became a shadow of its former self until, in the middle of the nineteenth century, it found a new role as the basis for photography.

Light-sensitive silver salts

In the hot New York summer of 1973, I spent weekend afternoons in Washington Square Park taking photographs. In the centre of the park an array of chess tables were always crowded with old men, faces staring intently at the black and white chequered boards. As I pressed my shutter release, light reflected from this city scene streamed in through the open aperture and on to the silver halide film. Where photons struck, atoms of pure silver formed, recording an image of history.[28]

The idea of permanently fixing images was first conceived by Thomas Wedgwood, the son of Josiah Wedgwood, the renowned English potter, at the end of the eighteenth century. Josiah had used a camera obscura, which projects an image of its surroundings on to a screen, to draw images more rapidly and accurately on to his pottery. Wedgwood wondered how he might be able to give these images permanence and set about experimenting with silver nitrate, whose light-sensitive properties had been known since an accidental discovery by a German university professor in 1725. Johann Schulze was investigating the properties of a nitric acid and chalk solution (which also happened to contain some silver) at the University of Altdorf, near Nuremberg. He was working near a window and, since it was a sunny day, light streamed into the clear jar in which he had placed the solution. Suddenly he noticed that the part of the mixture facing the window had turned purple, while that facing inwards was still white. Perhaps, he thought, it was the heat of the sun that was causing a chemical reaction in the solution. He tried the experiment again, but this time kept the jar in the dark. Nothing happened. Schulze realised it must have been the action of the sunlight transforming the mixture and, after further investigation, concluded that silver was the vital element in this reaction.[29] Wedgwood used this discovery. He coated sheets of paper with silver nitrate solution and then placed objects on top of them. By exposing the sheets to sunlight, he created silhouettes of the objects but sunlight would gradually blacken the remaining silver nitrate and the image would disappear.[30] Frustrated and suffering from ill health, he stopped his experiments. The invention of photography, which created permanent images, would have to wait another thirty years.

In October 1833, William Henry Fox Talbot, another Englishman, was on honeymoon on the shores of Lake Como. He was trying to sketch the landscape using a camera lucida, another draughtsman's aid which uses a prism to superimpose the image of the landscape over the paper. He was having little luck. When he removed his eye from the prism, he 'found that the faithless pencil had only left traces on the paper melancholy to behold'.[31] Fox Talbot wanted to find a better technique and so on his return home, apparently unaware of Wedgwood's experiments, he began creating silhouettes of leaves, lace and other flat objects using paper coated with silver nitrate. But just as Wedgwood had found, he could not stop the images from fading. One day he noticed that the edges of the paper, which had been treated with only a small amount of salt solution (used as a base for the silver nitrate coating) were often much more light sensitive. By using a strong salt solution he saw that he could make the paper much less light sensitive. And so he realised that if he soaked the exposed paper in a strong salt solution, his images would be fixed in place. But that was only half the battle. The images from a camera obscura were too faint to make an impression. He set about experimenting to improve the light sensitivity of the paper and, by using lenses, which focused the light on to a smaller area, was able to create images the size of postage stamps. The images 'might be supposed to be the work of some Lilliputian artist', wrote Fox Talbot.[32] He had got further than Schulze and Wedgwood, but like them failed to comprehend the significance of his discovery; he put it to one side to pursue other interests.

Meanwhile, across the English Channel, a French theatre designer was experimenting with silver's light-sensitive properties, using silver-coated copper plates and iodine. In spring 1835, Louis-Jacques-Mandé Daguerre put an exposed but unsuccessful plate away in a cupboard to re-polish and use again. Returning a few days later he saw that an image had miraculously appeared. It turned out that mercury vapours from a bottle in the cupboard had developed the 'latent' image. When the plate had initially been exposed, silver atoms had created a hidden image, but the number of atoms was too small to make it visible. The mercury vapour amalgamated with the silver atoms in the latent image, making it visible. Using this process, Daguerre was able to reduce the exposure time to only twenty minutes, short enough to capture sharp images of unmoving objects. On 19 August 1839 the Académie des Sciences in Paris announced the invention of the daguerreotype.[33] When the competitive Fox Talbot heard of this, he quickly created his own method of latent-image production using a new type of paper with an exposure time of around a minute.[34] He called the new process the calotype.[35]

Almost forty years had passed since Thomas Wedgwood's first attempt to capture images using the light-sensitive properties of silver. There were now two successful and competing commercial products; Fox Talbot and Daguerre were each soon battling to convince the public of the advantages of their particular photographic method. Fox Talbot had some initial difficulty in persuading the public of the advantages of his process.[36] It did not produce as detailed an image as the daguerreotype but it could produce multiple images from the same negative whereas the daguerreotype could not, limiting it to expensive one-off projects such as portraiture for the wealthy. Fox Talbot's invention won out and it laid the foundations of photography, built on silver, for the next two centuries.

The proliferation of images

I took my first photographs in Singapore when I was four years old. It was the Coronation of Queen Elizabeth II, and Singapore, once a Crown colony of the British Empire, was heaving with what we now call street parties. With my Kodak Brownie, I took photographs of my family, my friends and everything I could see for as long as I had unexposed film. My Brownie camera was eventually replaced by a better version and then a whole series of cameras up to today when a set of great Leica cameras are my tools. These machines continue to help me record instants of action and momentary scenes, with occasional success. Kodak released the first Brownie camera, named after a set of popular cartoon characters, in 1900. The simple cardboard box with a roll film was portable and easy to use. It was an instant success and especially popular with children. Costing five shillings (around a quarter of an average week's wages), it was affordable to almost anyone. Its creation was made possible by the development of gelatin silver bromide-coated plates. These plates could be exposed, stored and, critically, only developed later in a laboratory. They were thus called 'dry' plates. 'Wet' collodion plates had to be coated, exposed and developed in a short space of time, requiring photographers to carry a heavy and cumbersome portable laboratory everywhere with them.

George Eastman, the founder of Kodak, took his first lessons in the wet-plate process in 1877 and soon discovered its impracticalities.[37] 'My layout, which included only the essentials, had in it a camera about the size of a soap box, a tripod, which was strong and heavy enough to support a bungalow, a big plate holder, a dark-tent, a nitrate bath, and a container for water.'[38] He had read about new developments in dry plates and decided to experiment himself. By 1879 he had created his own plates, which could be stored for much longer before developing than other products on the market. Eastman was still not satisfied: his glass plates were heavy, fragile and expensive. As Henry Ford would later do with the Model T automobile, he sought to put photography in the hands of the wider public.

He wanted 'to make the camera as convenient as the pencil', but to do this he needed to invent a cheaper and simpler photographic process.[39] He developed 'Eastman Negative Paper', which used rolls of paper, rather than glass, to capture the image. A roll of Eastman film could fit into a small black box, around which he designed a camera reduced to its essential components: the lens was fixed and the shutter was released by pulling a string.[40]

The Kodak camera first went on sale in 1888, costing $25 ($600 today). Its popularity enabled Eastman to produce cameras on a mass scale and so also reduce the price of each unit. In 1896, the 100,000th Kodak camera was produced; by then each one cost only $5 ($120 today). Kodak cameras appealed to a huge market because of their simplicity and affordability. Eastman had done what great inventors of consumer appliances have always done. He allowed the user (photographer) to focus on his purpose (taking photographs) while making the technical functions (developing and printing) unseen.[41] Eastman's cameras were sold ready loaded with a one-hundred-exposure film that was returned to the factory when finished. A few days later the processed negatives would be delivered as mounted prints. As Kodak's slogan went: 'You press the button, we do the rest.'[42] By the end of the following year, his factory was processing about seven hundred films a day. Now anyone could be a photographer. Kodak cameras enabled individuals to keep a record of their personal history and to fix memories of their most treasured moments. Photographs were now seen everywhere. They became not only common and disposable items, but also documents which recorded world-changing events, producing images that have shaped generations.

The decisive moment

Some images define cruelty, brutality and intolerance. The image of a Vietcong prisoner with a gun held to his temple, in the moment before execution, is, for me, synonymous with the Vietnam War. The photograph of crowds of civilians waiting for food in a sprawling Rwandan refugee camp will for ever remind me of the genocide which took place there. And the image of emaciated inmates pressed against the barbed-wire fence of the Auschwitz concentration camp describes the fate of so many members of my family in the Holocaust. Other images show much happier times, however. The photographs that I have taken from my windows overlooking the Grand Canal try to capture the moods of Venice as the seasons change. The photograph I made and which sits opposite my desk is of fascinated people in Santa Maria Novella, Florence. They are looking at the fresco, which demonstrates perspective, painted by Masaccio. The poetic images of the newly crowned Queen Elizabeth II, with painted backdrops of Westminster Abbey, by Cecil Beaton were designed to give joy to a Great Britain recovering from the Second World War. We learn of and remember momentous world events through photographs. Iconic images create shared experiences; they define generations.[43]

By the outbreak of the Second World War, more portable cameras with film that was more sensitive to light enabled the photographer to follow soldiers into battle, in contrast to earlier war photographers such as Mathew Brady and Roger Fenton, whose nineteenth-century equipment required their subjects to remain still. War photographer Robert Capa was with the second wave of troops in the Allied invasion of Normandy in 1944. Waist-deep in the English Channel, machine-gun bullets tearing up the water around him, Capa took some of the defining images of the D-Day landing. 'If your pictures aren't good enough, you aren't close enough,' Capa would say.[44] Photographs were able to capture fleeting images of the world that, without silver, would exist only as a faded memory. 'There is a creative fraction of a second when you are taking a picture,' explained Henri Cartier-Bresson, the prolific and masterful photographer, in 1957. 'Your eye must see a composition or an expression that life itself offers you, and you must know with intuition when to click the camera.'[45]

Making that click at the right moment created images that revealed how civilians had become tangled up in the Vietnam War. No longer was the enemy a faceless monster, but a suffering human being to whom viewers could relate. By revealing the common humanity on the other side of the world, photography awoke the global social conscience. Eddie Adams' 1968 photograph 'Saigon Execution', alluded to above, provoked

protests against what appeared to be the unthinking murder of an innocent Vietnamese citizen. His photograph also demonstrates the power
that photography has to distort actual events. Adams explained: 'The general killed the Viet Cong; I killed the general with my camera. Still photographs are the most powerful weapon in the world. People believe them;
but photographs do lie, even without manipulation. They are only half-
truths ... What the photograph didn't say was, "What would you do if
you were the general at that time and place on that hot day, and you caught
the so-called bad guy after he blew away one, two or three American
people?"'[46]

While photography comes close to a true representation of reality, it
is by no means a wholly reliable source. Images can be and are deceptive. Joseph Stalin famously had photographs altered as part of his purge
of Party enemies; commissars vanished from photographs as they were
selectively purged. The Chinese Communist Party, too, has made ample
use of distortion created in the darkroom.

From trusted sources, however, come gripping images of reality. In
1994, photojournalist Sebastião Salgado spent three days at the Kibeho
refugee camp in south-west Rwanda. Hundreds of refugees arrived each
day, fleeing rape and murder at the hands of militias. Salgado was shocked
by what he saw: 'The scale of the horror seemed to numb people to the very
idea of death. I saw one man walking with a bundle in his arms, chatting to
another man. When he arrived at the mass grave, he tossed the inert body
of his baby on to the pile and walked away, still chatting.'[47] As an economist, Salgado wanted to use images to show the reality behind the statistics
describing the world's poor. Showing is more powerful than telling: vivid
images of human suffering move beyond simple political messages and
invoke wounding, personally touching detail. In doing so, photographs
compel the viewer to act.

For me, silver has other ways to store precious memories beyond
photography. On the coffee table in my study sit three Persian silver
boxes, given to my parents as parting gifts when they left Iran in the 1960s.
Their elaborate designs were fashioned by repoussage, the technique of
hammering a metal from the back to create a pattern in low relief. They
bring back vivid memories of my time in Iran: the roar of gas flares,

the night-time glow from the oil fields, and, of course, our visits to the silversmith, who would sit cross-legged in a hut to ply his trade. In ornaments, decoration and jewellery, silver has always preserved memories. But it also continues to store value. During the 1970s, when silver photography was thriving, two brothers seeking a safe haven for their extraordinary wealth caused the price of silver to rise to a height never seen before or since.

Silver's last gasp

Bunker Hunt was first and foremost an oil man. He had inherited his interest in oil from his father, H. L. Hunt, the richest oil producer in Texas. H. L.'s first fortune was made playing poker in the aptly named oil town of El Dorado (and where, in 1926, Bunker was born). With his winnings he invested in a small oil lease and was lucky enough to strike oil with his first well. Though the well soon ran dry, he invested in several more successful wells, moving out into Florida and then back to Texas. Here he was the largest independent operator in the vast East Texas oilfield. By the time Bunker Hunt began to follow in his father's footsteps, there were few opportunities for oil exploration in the US and so he looked abroad to Libya. The Libyan government was at the time (during the reign of King Idris) selling concessions to foreigners to drill for oil. The race was on to buy up the most prospective areas. Hunt, an independent up against the well-equipped and well-financed international oil companies, won a concession in the eastern province of Cyrenaica, close to Libya's border with Egypt. Another concession, a T-shaped plot ('T for Texas' Bunker would say) called Block 65, much further inland in the Sahara Desert, seemed less promising, but Hunt was able to buy it on the cheap.[48]

He had little initial luck and was fast running out of money when, in 1960, he struck a last-ditch deal with BP. He agreed that BP could drill in Block 65 for a 50:50 split of any oil found. The gamble paid off; in November 1961, BP struck the huge Sarir oilfield, estimated to hold between eight and ten billion barrels, at least half of which was thought to be recoverable.[49] Even with the low oil prices of the 1960s, Hunt's half-share would be worth an impressive $4 billion which, along with the fortune inherited from his father, made him the richest man in the world. But the fortunes of both BP and Bunker Hunt were soon to change when, in 1969, a coup brought Colonel Muammar Qaddafi to power. British troops had taken a role as a guardian in the Persian Gulf since 1835, but in December 1971 they were about to withdraw. The day before their departure, Iran inserted itself in the power vacuum by taking control of three small strategic islands. The British did not resist the occupation, provoking a backlash from the rest of the Arab world. BP was at the time majority owned by the British government. So, as an act of retribution against Britain, Qaddafi nationalised BP's Libyan oil operations.[50] Hunt was still able to produce his share of the Sarir oilfield, some 200,000 barrels a day, but the next year Qaddafi ordered all the oil companies to give Libya a 51 per cent share. Unlike many other companies, Hunt refused. In April 1973, his Libyan operations were nationalised to give, as Qaddafi expressed it, 'a slap on America's cool, arrogant face.'[51] Bunker Hunt had become the richest man in the world and then gambled and lost almost everything. It would take a daring business mind to gamble again, but Hunt did just that. Oil no longer seemed a safe investment and so he invested in real estate and horses, which were his hobby.[52] He also turned his attention to silver.

In the 1970s, silver seemed an increasingly safe store of value. High inflation was eroding the value of money and precious metals looked like a good investment. Silver, which was selling at a low of $1.50 an ounce, seemed more of a bargain than gold. Hunt's view was that 'just about anything you buy, rather than paper, is better. You're bound to come ahead in the long pull. If you don't like gold, use silver, or diamonds, or copper, but something. Any damn fool can run a printing press.'[53] Bunker Hunt's brother, William Herbert Hunt, was also interested in investing in silver.[54] He had swallowed the assertions in *Silver Profits in the Seventies* by Jerome Smith which declared that 'within our lifetime, and perhaps within the decade, silver could become more valuable than gold'.[55] Herbert Hunt explained his logic for investing: 'My analysis of the recent economic history of the United States has led me to believe that the wisest investment is one which is protected from inflation. In my opinion natural resources meet this criterion, and, to that end, I have invested in oil, gas, coal, and precious metals, including silver.'[56] In the early 1970s, the ratio of the value of gold to silver was around 20:1. Bunker believed that this ratio would eventually return to its historical level of around 10:1, which, according to his logic, meant that, even if gold remained at its then prevailing price, silver would have to double in value.

The Hunts began buying millions and millions of ounces of silver. By early 1974, they had contracts totalling 55 million ounces, around 8 per cent of the world's supply. Unusually, they were taking physical delivery of their contracts. They were looking for more than short-term profits from speculation; they wanted to create a permanent and literal store of value. But by taking ownership of silver, they began to reduce its physical supply and drive its price higher and higher. Worried about the US government interfering, as they had done with gold in the 1930s, the Hunts decided to move a good portion of their silver to vaults in Europe.[57] At a 2,500-acre ranch to the east of Dallas they recruited a dozen cowboys by holding a shooting match to find out who were the best marksmen. The winners got to ride with guns on three unmarked 707s, protecting the Hunts' silver hoard on a flight to Europe. Arriving at New York and Chicago in the middle of the night, they loaded 40 million ounces of silver from armoured trucks on to the aeroplanes. The cowboys climbed aboard and then flew to Zurich, where the bullion was delivered to six secret storage locations.

But storing silver was very costly; to stop their wealth being eroded, the Hunts needed prices to increase, not just hold steady. Fortunately a combination of the Hunts' bullish behaviour on silver, new big buyers in the market and the general trend in commodity prices during the high inflation of the 1970s led to just that. Between 1970 and 1973 the price of silver doubled, rising to $3 an ounce, and by autumn 1979 it had reached $8 an ounce. Then suddenly in September it leapt to over $16 an ounce. As a result of their buying, the Hunts faced investigation by the Commodities Futures Trading Commission (CFTC). The CFTC was growing worried that a silver squeeze (in which insufficient bullion is available to meet future contracts) could be on the horizon, and so they asked the Hunts to sell some of their silver. The Hunts abhorred federal interference and, characteristically, declined.

Instead they kept on buying. By the end of that year, the Hunts controlled

90 million ounces of bullion in the US, with another 40 million hidden in Europe and another 90 million through futures contracts. On the last day of 1979 the price had reached $34.45 an ounce. By the beginning of 1980, the CFTC became increasingly worried and determined that the Hunts were 'too large relative to the size of the US and world silver markets'.[58] Backed by the CTFC, the trading market announced new limits restricting traders to no more than 10 million ounces' worth of future contracts. But still the price kept rising and the Hunts kept buying silver outright. On 17 January the price reached a new high of $50 an ounce. At this point, the Hunts had silver holdings worth nearly $4.5 billion and their unrealised profit on this was around $3.5 billion.[59] The authorities reacted four days later by closing down the silver futures market, with disastrous consequences for the Hunts.[60] The price of silver plummeted to $34 an ounce, levelling out in this region for the rest of the month. The price was partly kept down by people selling scrap silver, digging out old silver dinner place settings to melt into bullion. 'Why would anyone want to sell silver to get dollars?' asked Bunker. 'I guess they got tired of polishing it.'[61] The Hunts kept taking delivery of their contracts, now holding stock of over 155 million ounces. They had faith in the long-term price of silver and tried to turn the price decline around. By 14 March the price of silver had fallen to $21 an ounce. On 25 March, the Hunts realised they no longer had the resources to keep making margin calls. Bunker sent a message to Herbert: 'Shut it down.' Their brokerage firm called asking for a further payment of $135 million to maintain the futures contracts. Herbert told them that they could not pay and that they would have to begin selling; as they did, panic took over the market. On Thursday 27 March, 'Silver Thursday', the silver market collapsed, falling to $10.80 an ounce. The Dow Jones dropped twenty-five points to its lowest level in five years. To pay their debts, the Hunts had to keep selling their silver, depressing the price further. And they also had to pledge their personal possessions, including 'thousands of ancient coins from the 3rd century BC, sixteenth-century antiques, Greek and Roman statuettes of bronze and silver ... a Rolex watch, and a Mercedes-Benz automobile'.[62]

The Hunts had made billions in profit on their investment, and now had lost it all. Bunker Hunt was twice the world's richest private individual (once by oil, once by silver) and twice had lost his fortune.[63] He still believed the price of silver would rise again, predicting it would one day reach $125 an ounce. But silver had probably had its last gasp. The Hunts' actions in the silver market put them in the media spotlight, where they were portrayed as greedy and conniving. They were hounded by the press, but as Bunker Hunt told a friend: 'At least I know they're using a little bit of silver every time they take my picture.'[64]

Silver where?

Since the first Kodak came on to the market in 1888, cameras and silver halide film had spread throughout the world. In 1980 silver images were more prolific than ever before. However, only thirty years later, this would all change. The 2012 Oscar's ceremony took place, unusually, without a sponsor. Kodak, whose name had been associated with the awards since 2000, had gone bankrupt. The pioneers of affordable cameras and film had failed to keep up with the shift to digital photography. On the red carpet, the press pack did not even use the tiniest bit of silver each time they took a photograph. In 2012, the silicon CCD chip was the medium of choice for recording images. Silver changed so much of our world, not only as a store of value but also as the essential ingredient in a technology allowing us to capture images of history. Silver-halide photography had a remarkable 150-year impact on the history of human invention, but that had now largely come to an end. While our reverence for gold, 'the sweat of the Sun', remains the same today as it did three millennia ago, we shed a tear for the 'tears of the Moon' and its gradual demise as an element which changed the world.[65]

URANIUM

The bomb

28 October 2011: I am standing outside the Peace Memorial Museum in Hiroshima, Japan. A plaque, 100 metres to the north-east of the museum, marks the hypocentre of the atomic bomb dropped on Hiroshima at 8.15 a.m., 6 August 1945.[1] Seventy thousand people died as a direct result of the explosion and many more died later from their burns and radiation sickness. The entire city was obliterated.[2]

The Peace Park's central boulevard points me towards the Atomic Bomb Dome, next door to the 'T'-shaped Aioi Bridge which served as the aiming point for the crew that dropped the bomb. The Dome was one of the few buildings left standing after the blast and the fires which swept through the city. The hollow, twisted structure has been kept in its bombed state as a lasting reminder of the destructive potential of atomic and nuclear weapons and as a memorial to those who lost their lives. But around the Atomic Bomb Dome and symbolically flat Peace Park, the city has been reborn. By 1955 the population of the city had surpassed the level reached before the bomb was dropped. Skyscrapers now dominate the skyline and the starkly contrasting Dome seems to have become ever smaller through the decades of Japan's post-war economic miracle.

In front of me is a sea of matching yellow, blue and green hats. Each day hundreds of school children visit the Peace Park. Some laugh and play games while others listen intently to their teachers. The Peace Bell frequently chimes out across the park as small groups take it in turns to swing the wooden ringer. They are all here for one reason: to learn about the tragic events of the day the bomb was dropped and why it must never happen again.

On display inside the museum is a collection of pictures made by Hiroshima survivors.[3] The images were produced decades after the bomb, but memories of that day are clearly still sharp. The individual hurt and terror is magnified as I move from picture to picture. The collection forms a partial documentary of human suffering. Drawn by the hands of survivors and registering what had been burnt in their minds, the images make a greater impact than that of any photograph. Many did not get to tell their story: a burial mound in the park still holds the cremated remains of thousands of unnamed victims.

The simplest images are the most affecting. In the middle of one blank white page sits a coarse black ball, limbs barely discernible. The flash of the atomic bomb was so quick that there was no time for any cooling to take place; asphalt boiled and skin simply burnt. The witness to this event continues: 'The whole body was so deeply charred that the gender was unrecognizable – yet the person was weakly writhing. I had to avert my eyes from the unbearable sight, but it entrenched itself in my memory for the rest of my life.'[4]

The Hiroshima survivors cannot forget; we must ensure that we do not forget.

At a dinner several weeks earlier, I had heard George Shultz, the former US Secretary of State, say that the abolition of nuclear weapons was his essential motivation.[5] At the time, I had not fully understood his drive, but as I stood in the Peace Park reflecting on the events of sixty-six years ago, it became clear to me. I felt the same when I had visited the US Holocaust Memorial Museum in Washington, DC, with my mother. Almost all her family died in the Second World War. She herself survived Auschwitz. Humans cannot be treated like vermin to be eradicated. The Holocaust was a planned programme of inhuman events for an evil reason; the Hiroshima bomb was a single dreadful event to stop a continuing inhuman war. I cried, much as I did when I saw my mother put a candle on the Auschwitz memorial. And I remembered my mother telling me off as we left the museum with a question and a statement. 'Why are you crying? It is just a museum. There is no noise and it does not smell.' And in that way she reminded me that there is no room for sentimentality, only clear-eyed realism.

Hiroshima most vividly illustrates that our use of the elements holds the potential for great evil as well as great good. The atomic bomb harnessed the incomparable energy of uranium; unleashed over Hiroshima in August 1945, it changed the world forever.[6]

Splitting the atom

Uranium was first mined as part of a greasy mineral called pitchblende, or 'bad luck rock'. During the Middle Ages, among the Cruel Mountains of the Kingdom of Bohemia, silver miners would thrust their pickaxes deep into the rock face only to pull out a clod of this unfortunate ore. When pitchblende was found it usually meant the end of a particular vein of silver and so the beginning of more back-breaking work digging a new mining shaft.

For centuries, uranium was discarded as waste, but, at the turn of the twentieth century, it became an ore of great scientific interest. It was from uranium salts that, in February 1896, Henri Becquerel discovered natural radiation. He placed some uranium salt on a photographic plate and a silhouette soon appeared. This effect was not considered unusual. Uranium had long been known to affect film in this way and Becquerel thought it to be a result of a chemical reaction triggered by sunlight. Because the next few days in Paris became overcast, Becquerel decided to suspend his experiments. He placed the photographic plates and uranium salts in a drawer. Returning days later he saw that the plates had mysteriously captured the same silhouettes. There had been no sun to make the images and so something else, intrinsic to the uranium, had made them. This he called radiation.

Fellow Parisian Marie Curie heard of Becquerel's discovery. To experiment further, Curie acquired a ton of pitchblende, still regarded as useless rock, free of charge from the Cruel Mountain region. This she used to obtain enough uranium salts to demonstrate that the amount of radiation emitted depends only on the mass of uranium. It did not matter whether it was solid or a powder, or exposed to heat or light. Her experiments led to her hypothesis that radiation comes from within the atom itself, rather than some chemical process between atoms. This radiation was the first sign of instability within the uranium atom.

Curie went on to discover two more radioactive elements, radium and polonium, and in 1903, along with her husband Pierre and Henri Becquerel, was awarded the Nobel Prize in Physics. Yet both Marie and Pierre Curie paid the price of their radioactive research. They succumbed to strange and, for Marie, fatal illnesses of a mysterious origin. Today we know that these resulted from radiation exposure, which at first was not known to be harmful.[7]

I first learnt of uranium's mysterious properties while reading Natural Sciences at Cambridge in the late 1960s. It was there, forty years earlier, that many of the pioneering experiments in the newly emerging field of nuclear physics had taken place. In 1920, Ernest Rutherford, 'the father of nuclear physics' and then the director of the Cavendish Laboratory, had hypothesised that the atom was made not only of negatively charged electrons and positively charged protons bound together by their opposite charges, but also of a neutral particle which he called a neutron.[8] Twelve years later, James Chadwick, assistant director of the Cavendish under Rutherford, proved the neutron's existence.[9] It would turn out to be the key to splitting uranium atoms.

By the time I arrived at Cambridge, the Cavendish Laboratory had lost much of its reputation for nuclear physics. One exception was the presence of Otto Frisch, a nuclear physicist who had played an active role in the Manhattan Project which produced the bomb dropped on Hiroshima. Years before, it was Frisch's interpretation of an unusual experimental result that, at a very basic level, provided the basis for the creation of the atomic bomb.

In 1938, Frisch was staying in Sweden over Christmas with his aunt and fellow physicist Lise Meitner. Meitner had been working at the Kaiser Wilhelm Institute for Chemistry in Dahlem, Germany, but had fled to Sweden earlier that year fearing persecution under the Third Reich. Her German colleagues, Otto Hahn and Fritz Strassmann, were keeping her updated on the progress of their research and in their latest correspondence told her of a very strange experimental result. When they fired neutrons at uranium atoms they detected signs of much lighter particles, almost half the weight of a uranium nucleus, in the debris.[10] It did not make sense: the neutrons did not have nearly enough energy to break off such large chunks of atom. They decided to go for a long walk to puzzle it out.

Walking through the frozen Swedish countryside with Meitner and his aunt, it occurred to Frisch that the uranium nucleus must be very unstable. Uranium atoms are so large that the glue holding the protons and neutrons together is only just enough to withstand the repulsive force of charged protons pushing against each other.[11] If a uranium nucleus absorbs a passing neutron, the energy is enough to destabilise the atom and split it. While he explained his reasoning to Meitner, both stopped and sat down on a tree trunk and began scribbling on bits of paper. They calculated that the uranium nucleus resembled 'a very wobbly, unstable drop, ready to divide itself at the slightest provocation'.[12] Frisch carried out further experiments to confirm this idea before he and Meitner submitted two papers to the scientific journal *Nature* in February 1939. In these, Frisch gave this newly observed event of splitting the name 'nuclear fission'.[13]

In nuclear fission a vast amount of energy is released. When a uranium atom splits it produces two smaller nuclei with a smaller total mass. The missing mass (m), which is about one-fifth of that of a proton, is converted into energy (E) according to Einstein's equation $E = mc^2$. Because the speed of light (c) is so fast (299,792,458 metres every second) a little mass makes a lot of energy. Frisch calculated that the splitting of a single uranium atom would be enough to make a grain of sand, itself containing around one hundred trillion atoms, visibly jump in the air.

But one uranium atom does not make a bomb. The explosion in Hiroshima resulted from the simultaneous splitting of some of the thousand trillion trillion atoms contained in one kilogram of uranium.[14] The splitting of a single uranium atom produces neutrons which, if travelling at the right speed, cause neighbouring uranium atoms to split as well. These produce more neutrons which in turn split more uranium atoms. This leads to an unstoppable chain reaction. Thus, uranium is literally regarded as the master of its own destruction.

Meitner and Frisch had unlocked the secret of splitting the atom. At the time, they did not understand that their discovery would enable the creation of a nuclear bomb which could be used to kill hundreds of thousands of people.

No explosion was observed in Meitner's laboratory. It turned out that the chain reaction could only be sustained with a specific type of uranium atom, a type which is rarely found in nature. Naturally occurring uranium is almost entirely made up of the isotope uranium-238.[15] However, in order to create a chain reaction, which results in a nuclear explosion, a high concentration of the fissionable isotope uranium-235 is needed. This is called 'enriched-uranium'.[16]

Building a city

At Columbia University in New York, Hungarian and Italian physicists Leó Szilárd and Enrico Fermi heard of Meitner and Frisch's research and began conducting their own experiments in nuclear fission. After moving to the University of Chicago they built the world's first nuclear reactor, Chicago Pile-1, in an old squash court under a football stadium at the university. This did not escape the attention of the Soviet intelligence agencies, but developments got somewhat lost in translation: the Soviets mistranslated 'squash court' as 'pumpkin field', mistaking the game of squash for the eponymous vegetable. The location of the experiment was irrelevant, though, in stark contrast to its results. From the large number of neutrons released in the fission reactions, Szilárd concluded that uranium would be able to sustain a chain reaction, and so possibly could be used to create a bomb. He later recalled: 'there was very little doubt in my mind that the world was headed for grief.'[17]

At the beginning of the Second World War Szilárd decided that the discovery should be urgently brought to the attention of President Roosevelt.[18] He drafted a letter, which was also signed by Albert Einstein, explaining the possibility of this new type of bomb and warning that Germany could be pursuing research with this aim in mind. Roosevelt agreed to the formation of the Advisory Committee on Uranium and, later, the allocation of increased funding for uranium research. As the science behind atomic explosions became more concrete, and as the urgency of the matter increased with the Japanese attack on Pearl Harbor and America's entry into the war, the diverse nuclear research in the US was consolidated and directed towards the single aim of producing an atomic bomb under the umbrella of the Manhattan Project.

In September 1942, the US government approved the acquisition of over 200 square kilometres of land surrounding the small town of Oak Ridge, Tennessee, to create the Clinton Engineer Works. As one of three main sites in the Manhattan Project, it was tasked with producing the enriched uranium for an atomic bomb. The other two sites under the umbrella of the Manhattan Project were Hanford, another production site for bomb material, and Los Alamos, the 'mind centre' of the project. Secrecy was a top priority for all three of the sites. They did not exist on maps and were referred to only by their code names of X, Y and Z.

As soon as Major General Leslie Groves, the director of the Manhattan Project, saw the site he knew it was right. Hidden away in the middle of nowhere, Oak Ridge was perfectly positioned for the project's secrecy and security needs. Being far from the coast reduced the risk of enemy attack and nearby rivers provided a plentiful supply of water and hydroelectric power, vital for the colossal industrial effort about to be undertaken.

With characteristic efficiency, Groves evicted 1,000 families from the area, some with only two weeks' notice. They had no choice. The site had been chosen, and no one was going to get in the way of America building the bomb. It was 'child's play' according to one official from the US Army Corps of Engineers.[19]

Scale was everything for the Clinton Engineer Works. At its peak it employed 80,000 workers. The small town of Oak Ridge quickly grew to become the fifth largest city in the state of Tennessee and held a greater concentration of PhDs per capita than any other city in the country. Twelve thousand people worked in the K-25 uranium-processing building alone. At the time, it covered more area than any structure ever built. Inside K-25, uranium was enriched by passing it, in gaseous form, through a series of membranes. Lighter molecules pass through a fine membrane faster than heavier molecules, so that the percentage of uranium-235, which was lighter than uranium-238, gradually increased.[20]

The US did not know whether their prodigious experiment would work, but they had to try. Only an atomic bomb, harnessing the incomparably destructive energy of uranium, held the potential to win the war in an instant. They also understood that speed in this endeavour was essential. They feared that Germany might beat them to it and unleash the devastating power of uranium on them.

To the US, these uniquely dark circumstances justified the enormous expense of the project and the forced evictions at Oak Ridge. They also enabled it to enlist the world's brightest and best scientific minds at the Los Alamos site, the 'mind centre'. Robert Oppenheimer, the director at Los Alamos, was among the first to witness the success of the US's endeavours at the Trinity bomb test site on 16 July 1945. Later he recalled that, as he watched the bright atomic flash in the New Mexico desert, he was reminded of a line from Hindu scripture: 'Now, I am become Death, the destroyer of worlds.'[21]

*

Today, scientific challenges are more diverse and the solutions less clear. This was evident when I visited the site of the Clinton Engineer Works, now called Oak Ridge National Laboratory, in March 2009. The square industrial buildings sit out of place among Tennessee's rolling forest landscape and the expansive countryside belies the true size of the Laboratory. On a tour of the site I was shown the X-10 graphite reactor, the second nuclear reactor in the world, and now a National Historic Landmark. The K-25 uranium enrichment facility is currently being demolished.

Research priorities have long since moved on, and these were the reason for my visit. I was there in my role as a partner in a private equity firm, which at the time managed the world's largest renewable and alternative energy investment fund. I came to learn about the Laboratory's recent approaches to the production of biofuel from crops, such as grasses and trees, which are obviously not used for food. Biofuel can be made from the sugars contained in plant cellulose. But non-food crops also contain a lot of lignin, which forms strong bonds with these sugars, making them difficult to extract. A particular interest at the laboratory was poplar trees, which have a wide variation in many natural traits. Researchers were searching over 1,000 poplar tree varieties for traits that would produce the greatest amounts of sugar.[22] By producing economically competitive biofuels sources, the US had hoped to reduce further their dependency on foreign oil. Oak Ridge may have moved on from uranium, but once again it was working in the interests of national security.

*

On 25 July 1945, the last shipment of enriched uranium needed for the Hiroshima bomb left Oak Ridge, arriving at the Pacific island of Tinian two days later. Here the three-metre-long atomic bomb, called 'Little Boy', which was soon to be dropped on the city of Hiroshima, was assembled. From this point on, the science was frighteningly simple. Lumps of enriched uranium would be slammed together to form a critical mass, initiating an uncontrollable, runaway nuclear reaction.

To create a bomb that would destroy a city, another city had been created. In total two billion dollars were spent 'on the greatest scientific gamble in history', a gamble that ultimately paid off.[23] The Manhattan Project is a rare example of a government successfully picking winners, yet in this case the choice was clear: only one weapon held the potential to end a war in an instant. The battles in the laboratories were as instrumental as those of the air, land and sea in securing Allied victory. In dropping the bomb, humanity had unleashed 'the basic power of the universe'.[24] But the bomb also made us fearfully aware of our newly found capacity for self-destruction; in doing so it symbolised the beginning of the modern age.

Up and Atom!

'At the instant of fission, Captain Atom was not flesh, bone and blood at all ... The desiccated molecular skeleton was intact but a change, never known to man, had taken place! Nothing ... absolutely nothing ... was left to mark the existence of what had once been a huge missile! Nor was there a trace of the man inside!'[25]

Captain Atom, 'radioactive as pure uranium-235', was born inside the explosion of an atomic warhead in the March 1960 issue of *Space Adventures*, a popular American comic book. Through the late 1950s and early 1960s, I eagerly read these and other science fiction stories. The mysterious power of the atom was a gift for comic book writers. They created a whole array of thrilling superheroes who could harness atomic energy as a force for good against the evils of the world. At the time, the greatest global threat, at least as far as America was concerned, was all-out nuclear warfare with the Soviet Union. In Captain Atom's debut adventure, he saves the world from destruction by intercepting a communist nuclear missile. 'You, more than any other weapon, will serve as a deterrent of war!' exclaimed the illustration of President Eisenhower on Captain Atom's jubilant return to earth.

Ever since the Little Boy bomb brought the power of uranium into the public eye, the possibilities of the Atomic Age seemed endless. While some atomic superheroes used their powers 'to crush every evil influence in the world' (Atomic Man) and 'save mankind from itself' (Atomic Thunderbolt), others were more villainous.[26] Mister Atom, a power-crazy nuclear powered robot, was hell-bent on taking over the world.

The same choice confronted the real world. General Groves, the director of the Manhattan Project, and regarded by many people as one of the fathers of the atomic bomb, also sternly warned that we had to choose the 'right path': weapons leading to atomic holocaust or a bright utopian atomic future.[27] Humanity, it seemed, stood at the fork in the road to a new atomic world.

The same rhetoric was apparent in the real President Eisenhower's 1953 address to the UN General Assembly. This came to be known as the 'Atoms for Peace' speech. Spurred by the sudden growth in the nuclear weapons arsenals of both the US and the Soviet Union, Eisenhower called on the world to 'strip the military casing' of the atomic bomb and adapt atomic energy for the benefit of humankind. He wanted America to lead the way in reducing nuclear stockpiles and to open up dialogue between the world's great nuclear superpowers. Eisenhower pledged that the US would 'devote its entire heart and mind to finding the way by which the miraculous inventiveness of man shall not be dedicated to his death, but consecrated to his life'.

In this imagined atomic utopia, it was believed that the unlimited source of neutrons produced in the splitting of uranium atoms would enable us to produce artificially atoms of any type in the laboratory. Since the beginning of humanity, we have sought to understand and harness the basic constituents of matter; now we could apparently take control over the elements in a way that was far more potent than any of the ancient alchemists had imagined.

Just as miraculous were the predicted medical benefits of radiation. Radioactive elements, people believed, would soon put an end to cancer, the most feared of diseases: one cartoon depicted a skeleton labelled 'CANCER' fleeing lightning bolts of 'ATOMIC ENERGY'.[28] So, too, would these radioactive elements enable us to trace a whole host of diseases as they made their way through the human body. By understanding these pathways, we hoped to develop a medical toolkit that would give a long and healthy life to all. And tracing similar pathways in plants would unlock the secrets of photosynthesis, harnessing the power of the sun and enhancing food production.

Of all the benefits of splitting the uranium atom, the most obvious was that it was a simple, abundant energy source. That was only too apparent in the destruction wreaked by the Hiroshima bomb. By harnessing the power of the uranium nucleus, fuel crises would become a thing of the past: we would soon all be driving atomic cars, each running on their own mini nuclear power generator.

Before Captain Atom hit the newsstands in 1960, my mainstay was the weekly *Eagle*, the leading British boys' comic of the 1950s. I remember studying the intricate centrefold cutaways of atomic submarines and aircraft carriers. Another image illustrated an 'Atomic Locomotive', 'the shape of things to come', moving at breakneck speed, powered by the unlimited energy of uranium.[29] With no need for refuelling or stoking, nuclear power was seen as superior to oil and coal. Using uranium, we would be able to travel across land and underwater farther and faster than ever before.

It was even thought that atomic energy would give us complete control of our climate. Artificial suns would control the weather, even, as one writer suggested, being used to melt ice caps to create a warm and temperate global climate.

The new nuclear energy source was unlike anything encountered before. Comic books and futurist writers give us a sense of the awe that uranium inspired in the 1950s and 1960s. Uranium was not unique in its technological potential being exaggerated. It is one of a trio of 'post-war wonder elements', along with titanium and silicon, whose stories are also explored in this book. The hyperbole surrounding uranium in this period was, however, greater than for any other element. The extraordinary power of uranium was self-evident in the images of mushroom clouds over Hiroshima. The fantastic imaginings that followed only raised uranium to a greater height from which, in many of its applications, it would ultimately fall.

Atomic-powered transport was, for most uses, deemed impractical and unsafe, while the notion of artificially heating our climate using uranium now seems ridiculous and, in light of anthropogenic climate change, anachronistic. Radiation is an important medical tool, but has by no means cured cancer. However, in one industry excitement appears to have been justified. Nuclear power stations, generating electricity from the heat produced in nuclear fission, soon began to appear across the globe.

An ambivalent institution

On 17 October 1956, Queen Elizabeth II pressed the switch at the Calder Hall nuclear power station in Cumbria, UK. For the first time, uranium's inherent energy was delivered directly to homes on a commercial scale.[30] Standing in the shadow of Calder Hall's cooling towers, the Queen presented the occasion as a solution to the dangers of atomic energy 'which has proved itself to be such a terrifying weapon of destruction'.[31] By using nuclear energy 'for the common good of our community', Britain wanted to give the impression of leading the way to the peaceful uses of atomic energy.[32]

The sudden development of nuclear power in Britain came about by necessity rather than choice: the harsh winter of 1947 had led to a national fuel crisis. Over 90 per cent of Britain's energy needs were supplied by coal in the post-war years, and in 1948 demand growth began to outstrip new supply. Reserves were fast disappearing, a shock for a nation which had once been one of the world's greatest coal exporters. Industrial Britain had prospered on these once abundant reserves; and if she was to continue as a major global economic power, a new energy source was needed.

Oil was one possible contender. In July 1954, the Minister of Fuel and Power announced that coal-fired power stations would be supplemented by imported oil. But this was only regarded as a short-term expedient. Concerns over diminishing oil reserves, which ultimately proved to be unfounded, demanded a longer-term solution. To Lord Cherwell, Chairman of the Atomic Energy Council, the mathematics was simple: 'one pound of uranium equals 1,000 tons of coal.'[33]

At the time I was living in Iran, where my father worked for the Anglo-Iranian Oil Company in the Masjid-i-Suleiman oilfields. Amid the excitement of well fires and the fascination with the mysterious process of oil production, concerns over energy security were far from my mind. Only later, back in the UK and considering university, did the importance of the nuclear power industry become apparent to me. The UK was beginning to invest in its second generation of nuclear power stations and the fast-growing, highly technological nuclear industry needed to attract the best minds. University education was a rarity then. Less than 5 per cent of young people went to university, and the industry offered lucrative scholarships to attract them. To me, the nuclear industry had a real sense of modernity. It seemed then that they were building the future and so I applied for a scholarship from the UK Atomic Energy Authority.

In the end I accepted a scholarship from BP, but it was a tough decision to make. I was attracted by BP's international dimension: the challenges the company faced seemed of greater scope and complexity. However, throughout my career in the oil industry, nuclear power has persisted in the background, coming to the fore in times of concern over oil supplies or following industrial accidents.

We are in just such a period now. The risk of anthropogenically induced climate change has once again brought the nuclear power debate to the surface. Nuclear power could help meet our increasing global energy demands in a low-carbon economy, but the growth of nuclear power suffers from its continued association with nuclear explosions, one which, from the very beginning, was interwoven with the production of nuclear power at Calder Hall.

Although held up as a paragon of peaceful nuclear energy, behind the scenes Calder Hall was used to create enriched uranium for atomic bombs. It was an ambivalent institution. Following the US bombings of Hiroshima and, subsequently, Nagasaki, the British government, like most other governments at the time, wanted an atomic bomb. In October 1946, the Prime Minister, Clement Attlee, called a cabinet meeting to discuss uranium enrichment for a nuclear weapon. They were about to decide against this on grounds of cost when Ernest Bevin, the Foreign Secretary, intervened. He was adamant and in an act of sadly human competitive behaviour said: 'We've got to have this ... I don't mind it for myself, but I don't want any other Foreign Secretary of this country to be talked to or at by the Secretary of State in the United States as I have just had ... We've got to have this thing over here, whatever it costs ... We've got to have the bloody Union Jack on top of it.'[34]

The plutonium used in Britain's first atomic bomb was produced at the site where Calder Hall was soon to be built. Calder Hall's reactor design was chosen primarily for its ability to produce the plutonium needed to keep pace in the global nuclear arms race. Uranium-238 is converted into plutonium when it is irradiated by neutrons released from nuclear fission reactions taking place inside the reactor. Rather than waste the heat produced in this process, electricity generators were also incorporated into the reactor's design. In its early life there was a trade-off between plutonium and power production: maximising one would mean diminishing the other. And often it was electricity output that lost out in favour of the British government's desire for a growing nuclear weapons arsenal.[35]

In the 1960s, as the US and Soviet Union became the focus of the global arms race, the UK's need for nuclear weapons decreased. And so Calder Hall prioritised the generation of electricity over the production of weapons; Britain increasingly separated its military and civilian nuclear programmes. Between the Queen's opening of Calder Hall in 1956 and the start of 2011, global nuclear electricity generating capacity had expanded to over 440 reactors, providing about a seventh of global electricity capacity. The growth of capacity had been slowing since the building boom of the seventies and eighties, but many industry analysts were predicting a 'nuclear renaissance' in the coming decade. At the start of 2011, Britain was considering building ten new nuclear power stations; China was planning to quadruple its nuclear power capacity by 2015; even traditionally anti-nuclear Germany was extending the lifetime of existing nuclear reactors. The bright future of nuclear power symbolised by Calder Hall seemed to be an ever-increasing reality. A few months later a single event would change everything.

Nuclear fear

On 11 March 2011, the Tōhoku earthquake sent a tsunami speeding towards the north-east coast of Japan. Almost 16,000 people died in the disaster, the great majority drowning in the tsunami flood waters.

The Fukushima Dai-ichi nuclear power plant stood 180 kilometres

1. All eyes on Sir William
Bragg. Nature explained.
To children and grown-ups
alike, 29 December 1931.

GEORGII AGRICOLAE
DE RE METALLICA LIBRI XII· QVI-
bus Officia, Instrumenta, Machinæ, ac omnia deniq; ad Metalli-
cam spectantia, non modo luculentissimè describuntur, sed & per
effigies, suis locis insertas, adiunctis Latinis, Germanicisq; appel-
lationibus ita ob oculos ponuntur, ut clarius tradi non possint.

EIVSDEM

DE ANIMANTIBVS SVBTERRANEIS Liber, ab Autore re-
cognitus:cum Indicibus diuersis, quicquid in opere tractatum est,
pulchrè demonstrantibus.

FRO · BEN

MVSEVM
BRITAN
NICVM

BASILEAE M· D· LVI·

Cum Priuilegio Imperatoris in annos v.
& Galliarum Regis ad Sexennium.

2. A master and his
masterpiece: Agricola's *De re
metallica* (1556), Prince Henry's
annotated copy.

IRON

3. *Above*: Ironclads fight it out. US Civil War, 1862.

4. No warfare is fought without iron (and men made of iron). Krupp addressing Hitler and Mussolini. Krupp factory, Essen, 1937.

5. The mighty *Thunder Horse* astride its transport.

6. *Opposite*: The Bessemer process in action. Forging steel in Carnegie's steel works, 1886.

7. *Above*: Bessemer lives on today: pig iron is poured into a Bessemer Converter. ThyssenKrupp, 2012.

8. Bessemer and Carnegie caught in a corner of the Institute of Materials, Minerals and Mining, London, 2012.

9. *Below left*: Henry Clay Frick painted with possessions.

10. Financed by steel but made without it: Carnegie Hall in the year of its opening, 1891.

11. A thousand years old and on sticks: the 19-foot tall Iron Lion of Cangzhou.

12. An engineering feat of the day: the Flatiron building. Stieglitz's iconic photograph, 1902-3.

13. Ruffling skirts on a postcard, 1907.

14. Carbon spectacular! Watched from a gondola. Festa del Redentore, Venice.

15. *Above left*: The site of the blowout: Rig 20 at Naft Safid, Iran, 1951.

16. *Above right*: Symbols of London. Smoked out, 1958.

17. *Right*: Man-made pollution hangs over a man-made lake. Kunming Lake, Beijing, 2008.

18. The carbon connection. Henry Ford, the inventor of his generation, flanked by me and his great-grandson, Bill.

19. *Below*: Oil collection in the sixteenth century: Agricola's bituminous spring. From Prince Henry's copy of *De re metallica*.

20. Exploring oil on stilts: Oily Rocks (Azerbaijan) on a 1971 Soviet postage stamp.

21. *Below*: A real seascape cut up by an unreal town: Oily Rocks from the air.

22. The *Exxon Valdez* disaster. Broken up in 2012, the ship no longer exists.

24. J. D. Rockefeller keeps an eye on things from the boardroom wall.

23. A platform engulfed in flames: Occidental Petroleum's *Piper Alpha* disaster in the North Sea, 1988.

25. From Russia with memory: my farewell to Vladimir Putin, 2007.

26. Drive-in movie? No such fun. Pumped-up panic in London, 1973.

27. A war engulfed in flames: burning oil wells in Operation Desert Storm, Kuwait, 1991.

28. BP + Amoco = Big Player. The birth of a Supermajor in the largest-ever industrial merger, 1998.

29. BP and PDVSA before Chavez in Venezuela: big negotiations. Big table. Big hair (me on the right-hand side). It was 1980!

30. Matters of state: visiting Patrick Manning, the Prime Minister of Trinidad and Tobago, 2007.

31. Putting gas on the right track: LNG trains in Trinidad.

32. Fracking a future of shale gas. In the midst of Lancashire's countryside.

33. Breaking Petroleum: the speech that broke ranks with Big Oil by addressing climate change, Stanford, 1997.

34. Solar, solar on the wall. Al Gore inspected them all, California, 1998.

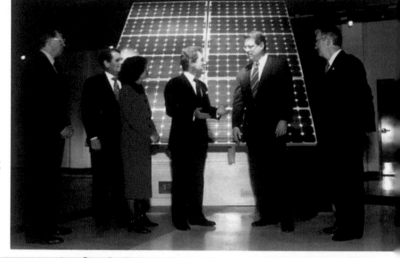

35. *Below*: 'Hasta la vista, climate change': terminating discussions with Tony Blair and Arnold Schwarzenegger, 2006.

GOLD

36. A myth forever kept afloat: the riches of El Dorado illustrated by its golden raft (Museo del Oro in Bogotá, Colombia).

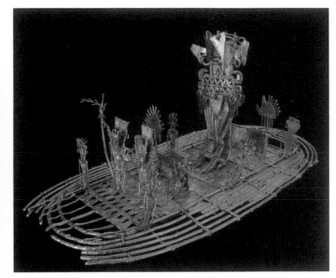

37. A gold *poporo* (South American storage jar) from my own collection.

38. Get gold! The Doge gives away gold on his inauguration. Brustolon's engraving (*c.* 1768) from a drawing by Canaletto.

39. Gold lasts! A Venetian ducat, the world's longest-lasting gold coin.

40. As rich as Croesus. Coinage of gold (and a phrase) inspired by a king.

41. Greed for gold: conquistadors execute Inca Emperor Atahualpa (sixteenth century).

42. A nugget that triggered a gold rush: Marshall's original nugget.

43. Found in California. Panned by Chinese workers. Gold rush, c. 1855.

44. Flip it, but don't clip it. Newton's invention lives on today.

45. And the biggest man-made hole on earth is . . . Bingham Canyon gold and copper mine, Uta

46. A farmer spotting a golden nugget in the mud was the origin of this gaping hole. Serra Pelada gold mine, Brazil.

SILVER

47. Silver smelting in
the sixteenth century: in
Agricola's *De re metallica*.

48. Watching over a silver
mountain: Madonna in Potosí.

49. The British calotype (1844) . . .

50. . . . and the French Daguerreotype (1838), the first photograph to show a human being

51. Say cheese: the beginnings of popular photography (*c.* 1910).

52. My attempt at a misty Venice from my window.

53. A five-year-old takes a shot at photography. My mother impatiently posing for my Brownie Box, Singapore, 1953.

54. 'Still photographs are the most powerful weapon in the world': a defining image of the Vietnam War, February 1968.

55. Auschwitz, 1945. A name and a horror no one must ever forget.

56. Personal memories preserved in Persian silver boxes.

57. The camera never lies. But does it? The case of the Vanishing Commissar.

58. The Hunt brothers: the last of the great silver moguls, 1980.

59. *Top*: The A-bomb dome dominates Hiroshima today: a legacy of an event that changed the world forever.

60. *Above*: A survivor of Hiroshima remembers. He was sixteen when the bomb was dropped.

61. *Right*: The fathers of nuclear horror: Oppenheimer and Groves by the remains of an atom bomb test, 1943.

62. Atoms for peace. HM Queen Elizabeth II opens Calder Hall nuclear power plant in 1956.

63. Captain Atom saves the world – again.

64. 'We've got to have the bloody Union Jack on top of it.' UK's nuclear deterrent, Blue Steel.

65. *Right*: Fukushima: the beginning of the end, 2011.

66. *Below*: Nuclear scientist Abdel Qadeer Khan at the inauguration of Pakistan's Ghauri-II missile, 1999.

67. *Above*: 'These are people, not pieces of paper' – Governor of Hiroshima, Hidehiko Yuzaki, 2012.

TITANIUM

68. Fit for any sky: Lockheed's supersonic
spy plane Blackbird, 1995.

69. Titan of the seas: a titanium Project 705
Soviet submarine.

70. A spectrum: Isaac Newton and
John Wickins, Cambridge, *c.* 1754.

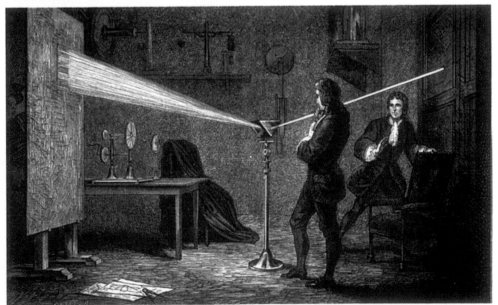

71. Titanium in art and architecture – the Guggenheim Museum, Bilbao.

72. Mining titanium in the midst of a beautiful lakescape. Rio Tinto's ilmenite mine, Lake Tio, Quebec.

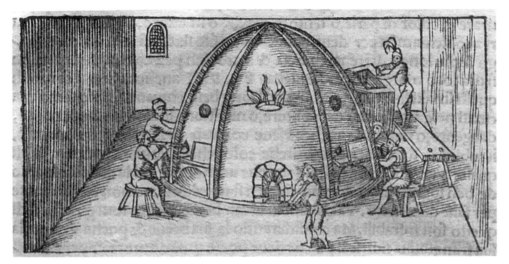

73. Glassmaking in Venice as depicted by Biringuccio in 1540 in the first printed book on metallurgy.

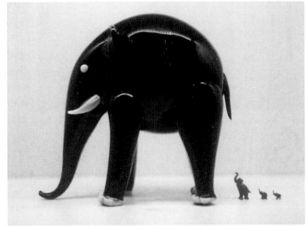

74. Breaking out of the herd: four of my glass elephants. Careful. They break! (2012)

75. Glass blowing in Murano. (2010)

76. *Left*: A virtual infinity: the Hall of Mirrors, Versailles, 2012.

77. *Above*: Glass petrol pump lights: silicon meets carbon.

78. *Below*: Paxton's magnificent Great Conservatory at Chatsworth. A dry-run for the mighty Crystal Palace.

79. A mighty palace
made not for royalty but
to showcase technology:
the Crystal Palace. The
two men on top are iron-
fitters, London, 1851.

80. 'If they had seen
what we see.' Mt Palomar
Observatory, 1959.

81. Sun in Palermo. Solar photovoltaic plant of AES Solar Italia.

82. Foresight of a genius. Insight of his works: Leonardo Da Vinci's solar machine.

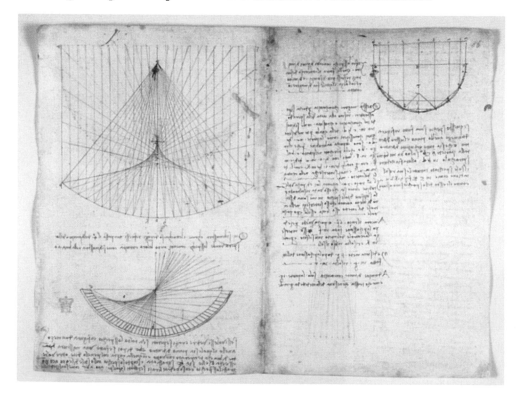

83. Computing in 1965. Overwhelming in power and size: the IBM 1130.

84. Fit for a museum of modern art: the first transistor of Nobel Prize-winning Shockley, Bardeen and Brattain.

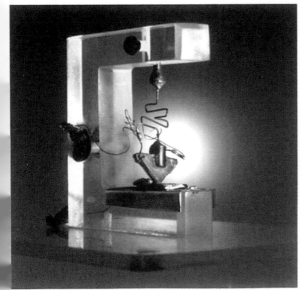

85. An Intel 22nm transistor. More than six million of these would fit into this full stop.

86. No Apple and Windows without Moore and Grove: the fathers of the modern silicon chip.

87. The big man behind silicon photonics: Mario Paniccia.

88. A microscopic view of a futuristic material: graphene.

away from the earthquake epicentre. It survived the initial magnitude 9.0 earthquake; this was one of the five most powerful shocks ever recorded anywhere on Earth. However, just under an hour later a towering 15-metre tsunami wave broke over the power station's flood defences. The complete loss of power and the subsequent failure of equipment led to a series of nuclear meltdowns and explosions, releasing radioactive material into the environment.

Fukushima reminded everyone that our control over the elements is far from perfect. An earthquake of this magnitude is extremely rare; but such events do happen and, for an element as potent as uranium, the consequences can be catastrophic.

When I visited Tokyo several months later, the country was still gripped by this catastrophe. The Fukushima power station had yet to be brought under complete control and newspaper headlines were dominated by all things nuclear; the fear was palpable. Concerns centred on how much radiation the plant had emitted and where this radiation had ultimately gone. Radiation levels across the country were published daily by the government. Newspapers in both English and Japanese carried maps of radiation levels. Everyone seemed to be an expert on sieverts, a measure of radiation, and the sale of Geiger counters, used to measure it, was booming. This was one of the few expanding markets in Japan's slowing economy. Once more, radiation pervaded the thoughts of the Japanese public.

Japan was the first and only nation to experience the terrifyingly acute effects of an atomic bomb and the persistent fear that exists in the aftermath. No one knew what the health effects of the bombings of Hiroshima and Nagasaki would be. Those unfortunate enough to be caught in the blast struggled to make sense of the mysterious illness befalling them, their friends and their family. It could be minutes, hours or days later when apparently healthy survivors would be struck down, the appearance of purple spots on the skin acting as an omen of death. Nausea would set in, followed by vomiting, bloody diarrhoea, fever and weakness. As a Hiroshima doctor, Michihiko Hachiya, wondered: 'Did the new weapon I had heard about throw off a poison gas or perhaps some deadly germ?'[36]

In Japan, survivors of the Hiroshima and Nagasaki bombings are known as 'hibakusha', meaning 'explosion-affected people'. They have for a long time suffered social stigma, being avoided by those around them for fear of contamination. Radiation continues to be misunderstood today. It has what Princeton academic Robert Socolow calls a high 'dread-to-risk-ratio': the perceived hazards of radiation are often much higher than the actual hazards.[37] These misconceptions stem from the inherently mysterious nature of radiation. Invisible yet penetrating, radiation is not easily perceived by our senses. You may be exposed without even realising it and, even after exposure, the effect on your health is very uncertain. You may suffer no adverse impacts; or you may contract cancer at some indeterminate point in the future. It is impossible to predict.

Individual exposure to radiation released from an atomic bomb and radiation released from an accident at a nuclear power station are of a greatly different magnitude. No one has yet died because of radiation released from Fukushima, and it is unlikely that anyone will die as a direct result.[38] The accident spread radiation over a very large area of Japan and in the months following it several potentially harmful concentrations of radiation were detected, even in Tokyo, 250 kilometres away. Radiation fallout was detected on school playgrounds, baseball stadiums and pedestrian walkways, although this could not always be attributed to Fukushima.[39] Except in a very few cases, radiation levels outside the immediate Fukushima evacuation zone were so low they are very unlikely to harm anyone. But the Japanese public are scared because, however small the risk, no one can be sure.

People were just as concerned after the 1986 disaster at the nuclear power plant at Chernobyl in Ukraine, where an explosion completely destroyed a reactor within a few seconds.[40] There was a large and tragic loss of life: thirty people died and another 106 were treated for acute radiation syndrome. Survivors became anxious and depressed not only because they were forced to leave their homes in the accident area but also because they had a lingering fear that radiation released from the reactor would cause cancer and other life-threatening illnesses.[41]

The more widespread effects of the Chernobyl disaster were, in reality, relatively small. While it is true that the accident did release large amounts of radioactive material, it was, however, heavily diluted as it spread across the whole of the northern hemisphere. Outside the accident area the levels of radiation were very low and caused few observable health effects.[42] Nonetheless, the reaction from the wider public was of alarm, disproportionate to the actual risks.[43] In Germany, where some of the highest radiation levels were detected, this alarm has had a lasting effect.

Public confidence in nuclear power and its safety has, unsurprisingly, fallen since the Fukushima disaster. Protests have been staged calling for Japan to abandon nuclear power outright. In the immediate aftermath of the disaster though, a poll by the independent US think-tank the Pew Research Center found that 46 per cent of the Japanese public would like nuclear power maintained at current levels, slightly more than thought it should be reduced.[44] As Sir David Warren, British Ambassador to Japan, explained to me when I met him at the British Embassy, the public were directing their anger towards government and companies rather more than towards nuclear power itself. Although the government reaction to the Tōhoku earthquake was much faster than its response to the giant 1995 Kobe earthquake, officials were still seen to be acting with a lack of clarity and decisiveness. There was far too little clear communication between the government, the public and the Tokyo Electric Power Company (TEPCO), the plant's owners. At one point Prime Minister Naoto Kan was overheard asking TEPCO executives: 'What the hell is going on?'[45] Many Japanese citizens believe, rightly or wrongly, that the plant's owners were hiding the true extent of the accident from the government and public alike.

TEPCO was an example of an old-fashioned corporation which only liked to put out good news. They have received the brunt of the criticism for failing to plan sufficiently for very large tsunamis and for making operating mistakes that may have increased the release of radiation into the environment. Of greater concern, Fukushima has revealed the government's inability to regulate adequately the nuclear industry. Numerous instances of bad practice have been uncovered since the disaster. Japan has long been governed by informal bonds of trust and, as a result, the regulator and the regulated were not sufficiently separated.[46] Fukushima is forcing a re-examination of these relationships. The Chernobyl disaster was another stark example of a lack of distinction between the regulator and the regulated and, as a result, poor application of regulation.

Chernobyl also epitomised mismanagement of a disaster by the nuclear industry. The first reaction of the Soviet government was to hide the accident from the world. Only when presented with irrefutable evidence of the accident did officials admit to the explosion. Credibility was damaged and trust evaporated. While the Fukushima accident in Japan was not mismanaged as badly or surrounded with as much secrecy, it has caused a similar breakdown of trust.

Yesterday's element

By May 2012, every nuclear reactor in Japan had been taken offline. Some of them had been permanently damaged by the Fukushima earthquake, but many were gradually shut down over the following year pending safety tests or routine maintenance. Many interpreted this as a political gesture rather than a pragmatic necessity. Japan remained without nuclear power until July 2012, when two reactors were restarted. When the Pew Center repeated its poll just over a year after the accident, it found that 70 per cent of respondents thought that Japan's use of nuclear power should be reduced, compared to just 20 per cent who thought it should be maintained at current levels.[47] As time has passed, the Japanese public appears to have become more comfortable with the idea of a nuclear-free future.

Japan had planned to grow its generation of electricity from nuclear reactors from 30 per cent in 2010 to 50 per cent by 2030. It is now planning to reduce this dependency as much as possible. In order for this to happen it will have to find new energy sources and reduce its energy consumption through efficiency and conservation. That is a great challenge for a nation which has few naturally occurring energy resources and is already one of the most energy efficient economies in the world.

Plans to grow nuclear power generation elsewhere in the world have also been cut. In the immediate aftermath of Fukushima, Germany's Chancellor Angela Merkel announced a three-month moratorium on its decision to extend the lifetime of existing nuclear plants. This led to the announcement in May 2011 that it would completely abandon all nuclear power by 2022. Many other countries placed plans on hold pending reviews of the safety of reactors.

So far, the statistics are in favour of nuclear power; it appears to have been safer than almost every other energy source.[48] Nonetheless, our fear of catastrophe and radiation means that we insist on continuously improving tough safety standards that are better than those of any other energy source. Those other energy sources are, of course, not absolutely safe; there will always be some risk to humans and the environment. But the thing they do not have to take into account is the dread of radiation.

Nuclear power could form one part of the solution to the problem of removing carbon from our energy sources. However, Fukushima may mark the beginning of its end, not just in Japan but across the world. Nuclear power generation was always an uneasy fit with the production of nuclear weapons. Many now see it as simply too expensive and cumbersome an option when compared with other forms of energy, including renewable energy sources. The time period between deciding to build a nuclear plant and starting to generate electricity is so long that possible changes in regulation and even demand during that period make returns on investment highly uncertain. In addition, the costs of processing uranium fuel and its radioactive by-products are higher than for other fuel sources. That all makes financing very difficult to obtain at any reasonable price. For reasons of commerce as well as public concern nuclear power is likely to face a bleak future.

People not paper

On the first anniversary of the Hiroshima bomb, survivors gathered at the destroyed Gokoku shrine to pray for those lives lost. The flags they carried read: 'World Peace begins in Hiroshima'.[49]

The dead were buried and the city was beginning the slow process of rebuilding itself from scratch. Hiroshima was home to almost 300,000 civilians at the time of the explosion, but during the Second World War it was also a city of considerable military importance, containing a communications centre, army storage facilities and military bases for about 43,000 soldiers. After it had been part of the brutal climax of the war, Hiroshima would be reinvented as a symbol for world peace and the total eradication of nuclear weapons.

While I was in Hiroshima in 2011, I met Hidehiko Yuzaki, Governor of

Hiroshima Prefecture, to discuss the Fukushima accident and the legacy of the Hiroshima atomic bomb. Hiroshima exists as a stark reminder of the destructive potential of nuclear weapons and the need to continue efforts towards their eradication.

The Nuclear Non-Proliferation Treaty is rooted in Eisenhower's 1953 'Atoms for Peace' speech, which subsequently led to the formation of the International Atomic Energy Authority. This organisation monitors nuclear activities and helps non-nuclear nations to develop peaceful nuclear power programmes. Coming into force in 1970, the Treaty is a legally binding agreement which prohibits any action by a signatory nation that helps the spread of nuclear weapons to non-nuclear nations. Those countries that already held nuclear weapons before the Treaty came into force must also 'pursue negotiations in good faith' towards complete abolition of nuclear weapons.[50]

If the Treaty had not existed it is estimated that at least thirty countries would have acquired nuclear weapons by now. Even so, there are nine nations today which between them hold an estimated 22,400 nuclear warheads.[51] Four of these nations – India, Israel, Pakistan and North Korea – stand outside the Treaty. Total eradication of nuclear weapons is still a distant ideal and many nations will go to great lengths to get them.

In October 2003, the cargo ship MV *BBC China* was intercepted by CIA agents on its way through the Mediterranean to Libya. Several shipping crates were hastily removed from the ship and taken to a high-security warehouse. The next day, as the CIA opened them, their suspicions were confirmed: they contained thousands of aluminium centrifuge components. Two days later American and British intelligence officers met Colonel Qaddafi in the Libyan Desert. Libya's subsequent surrender of nuclear weapons technology led to further discoveries about the network's operations and ultimately the arrest of many of those suspected to be involved, including Abdul Qadeer Khan.

Khan grew up in British India in the 1940s, before fleeing to Pakistan amid the growing conflict brought on by Partition. As a child, he had witnessed trains pulling into his home town station piled high with the dead bodies of Muslims, instilling in him a deep sense of national pride and hostility towards India. In December 1971, Pakistan faced a crushing military defeat by India. Coming to power as president following the conflict, Zulfikar Ali Bhutto saw nuclear weapons as the only way of countering his neighbour's greater military might.[52] A month into his presidency, Bhutto held a meeting with the country's top scientists and military personnel. He wanted a nuclear bomb within three years. 'Can you give it to me?' Bhutto asked. 'Oh, yes. Yes you can have it,' the group cheered.[53]

Their neighbour held similar aims and on 18 May 1974 the Prime Minister of India, Indira Gandhi, watched as the floor of the Rajasthan desert rose up into an iconic mushroom cloud. Operation Smiling Buddha had been successful and India was now officially a nuclear weapons nation. India's nuclear test flew in the face of the Nuclear Non-Proliferation Treaty, which had come into force four years earlier. India had never signed the Treaty, but its display of nuclear might came as a kick in the teeth to the West. India had developed its weapons using information declassified under the US Atoms for Peace programme, while the plutonium inside the warhead was produced from the spent fuel of a reactor supplied by Canada for producing electricity. Widespread international condemnation followed and sanctions were placed on India. But it was Pakistan that felt most threatened by India's now apparent capabilities. India's bomb, detonated less than 50 kilometres from Pakistan's border, was another show of force towards its neighbour.

India's bomb seems to have been the spark for Khan's involvement in the Pakistan project. He had immigrated to Europe in the 1960s to study and he now worked for a nuclear fuel company called URENCO in the Netherlands, giving him access to the secret details of the uranium-enrichment process. Following Operation Smiling Buddha, he wrote to Bhutto saying he knew how to produce the fissile material needed for a bomb and offering his services. Suspicions grew that Khan was involved in the transfer of classified information, and in 1983 he was sentenced *in absentia* by a Dutch court to four years in prison for espionage. Khan denied the charges, and his lawyer vigorously defended the case on the grounds that the information Khan had obtained was freely available in any university library. The case ultimately collapsed, and Khan claimed vindication.

With Khan's help, Pakistan succeeded in building an atomic bomb. On 28 May 1998, they demonstrated their new might in Pakistan's Chagai Hills.

Pakistan's Prime Minister Nawaz Sharif had no choice; it was a matter of national pride. On the detonation of their bomb, Pakistan was jubilant, as they had at last matched India's display of military strength. Khan became a national hero.[54]

Khan's primary motivation for developing nuclear weapons was to support his home country geopolitically and against their apparently bellicose neighbour. Pakistan's nuclear capacity may indeed have made India think twice before provoking Pakistan. Yet the motivations of others who spread nuclear weapons capabilities to states such as Libya, North Korea, Iran and possibly beyond surely extended far beyond national pride, and were more likely rooted in greed and narcissism than political ideology. We still do not know for certain how far these proliferation networks might have spread, or whether any of their components are still active.

Nuclear non-proliferation efforts over the last forty years have undoubtedly constrained the number of states owning weapons. However, the actions of proliferation networks demonstrate that the spread has not been stopped altogether. This failure is in part due to the inevitable one-sided nature of the Nuclear Non-Proliferation Treaty. International law is made in the interests of those with the power to create and enforce it, and consequently a great fault line runs down the middle of the Treaty.

Nations are split into nuclear haves and have-nots. The five nations (US, UK, France, Russia and China) to acquire nuclear weapons before the Treaty came into force are granted nuclear hegemony over all others. Any nation with a nuclear power programme, but no nuclear weapons, must open itself up to weapons inspections. There is no obligation for nations holding weapons to reciprocate. Those with nuclear weapons are unwilling to give them up, while those without strive to own them.

I lived through the tense decades of the Cold War, during which time the existence of nuclear weapons held us in an uneasy peace. In earlier centuries, two great superpowers like the US and the Soviet Union would undoubtedly have engaged in war, but the terrible prospect of a thermo-nuclear Armageddon held them on the brink.

Global peace relied on a simple threat: should either side launch a nuclear attack, the other would unleash its entire nuclear arsenal. The result would be the complete annihilation of that nation. As the US

Secretary of Defense Robert McNamara described it: 'Technology has now circumscribed us all with a horizon of horror that could dwarf any catastrophe that has befallen man in his more than a million years on earth … Deterrence of nuclear aggression means the certainty of suicide to the aggressor, not merely to his military forces, but to his society as a whole.'[55]

That statement encapsulated the doctrine of mutually assured destruction (MAD), and created a single strategic imperative: to ensure its safety, strategists concluded, the US needed the ability to inflict total destruction on any aggressor, even after sustaining a nuclear strike. The US nuclear deterrent had to be large enough to withstand the full onslaught of the Soviet Union's arsenal, and still wipe them out. To use the jargon, they needed to retain 'second strike capability'.

Soviet strategists reached the same conclusion, and so each side set about building a vast nuclear arsenal. For any increase made by the enemy, the arsenal had to grow further if it was to withstand a strike by the enemy. The great arms race of the twentieth century was on. By 1982, each side had more than 10,000 strategic warheads. They were spread around the globe to reduce the possibility they could be destroyed in a first strike: on inter-continental ballistic missiles hidden in reinforced concrete silos, on nuclear submarines deep under the seas, and on fleets of planes constantly circling in the air. The primary aim on each side was not to convey aggression, but to provide an effective and credible deterrent: 'if you attack me, you will die too.'

In the Soviet Union, the logic of second-strike capability was taken to its extreme in the construction of a 'Dead Hand' system, wrapped in secrecy but rumoured still to exist today.[56] Should a US nuclear attack wipe out the Soviet leadership, the Dead Hand would be triggered and an automated nuclear response would be launched. Dead Hand would send a series of unarmed missiles flying across the Russian continent, broadcasting a radio code to thousands of armed missiles, firing up from silos across the nation to obliterate America.

The consequence of mutually assured destruction was perverse and terrifying: huge nuclear arsenals, able to destroy the world many times over, were built in order that they might never be used. It created constant dread and occasional terror, but it sustained a peace of sorts.

But the world today is different. The simple balance of mutually assured destruction between two superpowers has gone. Instead, we have a multitude of nuclear actors, whose motives are complex and upon whose rationality we cannot rely. Among these nuclear powers, nationalism and identity politics can prevent critical thinking about the blunt reality of nuclear weapons. For each additional player, the risk of disaster increases, whether a launch is triggered by misinformation, misjudgement or mechanical accident. Perhaps most frightening of all is the prospect that weapons could fall into the hands of terrorists, for whom death is no deterrent.

In harnessing uranium, and unleashing the first nuclear reaction over Hiroshima, we tapped the primordial energy source of the Universe. For sixty years since, annihilation has been prevented by the threat of mutually assured destruction, but we can no longer rely on that uneasy equilibrium. The probabilities of disaster are too great, and the damage too severe. If a nuclear bomb were dropped today, the destruction would be many times that unleashed at Hiroshima.

Walking among the groups of school children in the Hiroshima Peace Park and Museum, I realised more than ever the need for education, of both the young and old, about the stark realities of nuclear weapons. As I sat and talked to Governor Yuzaki, who is leading a renewed drive for nuclear non-proliferation, we agreed that we must be hopeful about a future free of nuclear weapons. Constructive political discourse at an international level is not easy, if possible at all. But it is worth the effort if we can reduce the risk of one more nuclear bomb being dropped. Total eradication may be unrealistic, but we have to try.

With new generations, fresh thinking from minds that do not remember the Cold War years may see nuclear weapons for what they are in their simplest form: a terrifying weapon of unparalleled destruction. As we spoke of writing papers and treaties, Governor Yuzaki looked out on his city and deftly summed up the issue at hand: 'These are people, not pieces of paper.'

TITANIUM

In October 1950, *Popular Science* magazine featured a 'new rival' that 'challenges aluminum and steel as a structural material for airplanes and rockets, guns and armor'. Strong, lightweight and corrosion resistant, titanium was presented as the wonder metal of the future.[1]

Titanium was discovered in 1791 by William Gregor, an English clergyman, mineralogist and chemist, when he isolated some 'black sand' from a river in the Manaccan valley in Cornwall. We now know this as the mineral ilmenite, an iron-titanium oxide, from which he produced an impure oxide of a new element that he called manaccanite. Four years later, Martin Klaproth, a German chemist, isolated titanium dioxide from titanium's other major ore, rutile. He called the new element titanium, after the gods of Greek mythology the Titans, who were imprisoned inside the Earth by their father, Uranus. Klaproth also discovered uranium; he chose abstract names for both elements as, at the time, their properties were not fully known.[2] Yet, coincidentally, Klaproth's name turned out to be apt: like the Titans, trapped inside the Earth, titanium is strongly bound in its ore and is very difficult to extract.

It was not until 1910 that the metallurgist Matthew Albert Hunter, working at the Rensselaer Polytechnic Institute outside New York, created a sample of pure metallic titanium. In doing so he revealed titanium's remarkable physical properties. It took until the 1940s, 150 years after titanium's original discovery, to develop a commercial process to extract titanium from its ore.

Now, as tensions mounted at the start of the Cold War, each side, the US and the Soviet Union, was desperate to establish a technological advantage that would give it superiority in the seas, skies and outer space. Titanium seemed a new miracle metal that could do just that. The First and Second World Wars were fought with iron and carbon; the Cold War would be fought with titanium and uranium.

Titanium made possible the most extreme of Cold War engineering, such as the supersonic spy plane the Lockheed Blackbird. Flying at three times the speed of sound, Blackbird aircraft could outrun the most advanced Soviet missile technology, bringing vital military intelligence back to US soil within hours. The Blackbird is an awe-inspiring work of engineering and is still the fastest air-breathing manned jet in the world.[3]

Supersonic Blackbird

'We'll fly at [27,000 metres] and jack up the speed to Mach 3 ... The higher and faster we fly the harder it will be to spot us, much less stop us,' explained Kelly Johnson, Vice-President of Advanced Development Projects at Lockheed aerospace company, to a group of engineers.[4]

In the 1950s, at the height of the Cold War, the US was desperate to know about Soviet military capabilities. Proposed satellite technology had severe limitations: orbits were fixed and too predictable for their paths to go unnoticed, while images taken from outer space were often blurred.

Kelly Johnson believed his spy plane was the only way to gather adequate military intelligence and also ensure the safety of the pilots onboard. The first spy planes developed during the Cold War were converted Second World War bombers that were slow and travelled at low altitude, making them vulnerable to attack from the Soviet Union. Lockheed's U-2 state-of-the-art spy plane of the late 1950s could travel at heights of 21 kilometres and speeds of up to 800 kilometres per hour, but the Soviet Union was investing heavily in advanced anti-spy plane weaponry that could attack the U-2. The US was mindful of the vulnerability of the U-2 to the Soviet's anti-spy technology and sought to develop a new spy plane that could go even higher and faster. Indeed, in 1960, just as work on the Blackbird had begun at Lockheed, a U-2 plane was shot down and the pilot, Gary Powers, captured by the KGB.

The Blackbird, flying four times faster and eight kilometres higher than the U-2, was the realisation of that American ambition. The plan was incredibly ambitious: the US air force wanted to build a plane that no longer just hid from Soviet missiles, but which could outpace any missile that could lock on to it. Aircraft had flown above Mach 3 before, but only for short bursts using afterburners. The Blackbird would cruise at this speed. It would fly whole missions on afterburners. But to succeed in building such a sophisticated piece of engineering, Lockheed's engineers had first to learn how to harness titanium.

In 1959, work on Johnson's Mach 3 aircraft began at Lockheed's design and engineering facility, the 'Skunk Works', named for the unbearable stench given off by a nearby plastics factory. The engineers soon realised that titanium was the only lightweight metal able to withstand the high temperatures created at Mach 3 flight; steel was just too heavy.

At a height of 27 kilometres the air is so thin that it is almost a vacuum. The temperature is a freezing minus 55 degrees centigrade. Even so, the Blackbird's nose, travelling faster than a rifle bullet, is heated by air friction to over 400 degrees centigrade.[5] Near the afterburner, the temperature is over 560 degrees. If it were not painted black, giving the aircraft its name, the temperature would be even higher.[6] The temperature is so extreme that the aircraft expands several inches during flight. The frame and fuel tank would only align correctly at high speeds so that when on the ground fuel leaked through gaps and on to the runway.

Over nine-tenths of the Blackbird's structural weight was made of titanium. At the start of the project, no one had worked with titanium on such a scale before or in such extreme conditions. Only one small company in the US, the Titanium Metals Corporation, milled titanium and the sheets they produced were of uneven quality. Moreover, they could not find enough titanium to build the plane. The CIA searched the globe and eventually sourced an exporter in the Soviet Union who was unaware they were aiding the creation of a spy plane to be used against them.

During the design testing and construction of the aircraft, over thirteen million titanium parts were manufactured. Engineers encountered many problems in the process. Slight impurities can turn titanium brittle and, at first, some components would shatter when dropped from waist height. Lines drawn on with a pen would quickly eat through thin titanium sheets;

cadmium-plated spanners caused bolts to drop out; most mysteriously of all, spot-welded panels produced in the summer would fall apart while those produced in winter would hold. Eventually the source of contamination was found to be chlorine that was added to water tanks at the Skunk Works in the summer to stop the growth of algae.

Solutions were found for these problems, but they were expensive. Engineers had to work in a meticulously clean environment, pickle each component in acid and weld in a nitrogen atmosphere. The aircraft's costs rapidly spiralled into hundreds of millions of dollars.

But it *was* built and, on 22 December 1964, the Blackbird made its maiden flight, completing a supersonic flyby down the runway of the airbase for Johnson's amusement. Lockheed had succeeded in creating the most incredible aircraft in human history. It remains today an example of what can be accomplished when human ingenuity is combined with the extraordinary properties of the Earth's elements.

More than an engineering marvel, the Blackbird was a functional tool of war. It quickly began to show its worth on its first operational mission during the Vietnam War. The US military base at Khe Sanh in South Vietnam was under siege by the North Vietnamese army, but the US was unable to find the truck park which was supplying the enemy with troops and ammunition. On 21 March 1968, the Blackbird flew a reconnaissance mission over the Demilitarised Zone between North and South Vietnam. The photographs revealed not only the suspected truck park, but also the placement of heavy artillery surrounding Khe Sanh. A few days later, the US launched air attacks against these targets, and within two weeks the siege had been lifted.

Having proved its worth in Vietnam, the Blackbird was put to use once again in October 1973 when Egyptian forces crossed the Suez Canal, instigating the Yom Kippur War. Israel was caught off guard by the sudden Arab attack and the US, which supported Israel in the conflict, feared that without adequate intelligence Israel could lose more ground. The Soviet Union, which supported the Arab forces, had repositioned its Cosmos satellite to provide information on Israeli troop positions; President Nixon ordered Blackbird to provide similar support to Israel.

The Blackbird flew from New York to the Arab-Israeli border, a distance of 9,000 kilometres, in a record time of five hours. Twenty-five minutes flying in restricted territory was all that was needed to photograph the battle lines below. By the next morning the images, showing the positions of the Arab forces, were on the desk of the Israeli general staff.

With titanium, the US controlled the skies during the Cold War. In space, too, titanium gave the Americans an advantage: it was used extensively in the Apollo and Mercury space programmes.[7] But on the other side of the Iron Curtain, the Soviet Union was also using the wonder metal titanium to rule the seas, building a new class of submarine that was smaller, faster and could dive to greater depths.

Soviet subs

'It must have an advanced design: new materials, a new power plant and a new weapons system – it must be superlative,' said Dr Georgi Sviatov, at the time a junior Soviet naval engineer, of the *K-162* submarine.[8] The Soviet Union sought to create a submarine that could pass quickly and undetected through hostile waters to attack the enemy.

The engineers considered steel and aluminium, but the superiority of titanium was clear. The strength-to-weight ratio of a metal is a crucial consideration in the construction of submarine hulls, which must be light so that the submarine is naturally buoyant, but which must also be able to face extreme water pressures. Titanium's superior ratio would enable Soviet submarines to dive to new depths. In addition, titanium is also corrosion-resistant, forming a thin layer of titanium dioxide on its surface which protects it against the harsh maritime environment. And, unlike iron, titanium is non-magnetic, reducing the likelihood of a submarine being detected and setting off magnetic mines.

As the US had done for the titanium Blackbird, the Soviet Union paid a premium for their high-tech hulls. The first titanium-hulled submarine, the *K-162*, was so expensive that most thought it would have been cheaper to make it out of gold; the submarine came to be known as the 'golden fish'.[9]

In 1983, the Soviet Union used titanium once again, this time to build the world's deepest diving submarine. The 400-foot-long *Komsomolets*, 'Member of the Young Communist League', was built with an inner titanium hull to operate at depths of up to a kilometre. The *Komsomolets* sank in April 1989 in the Norwegian Sea when a high-pressure air line burst and started a fire on board the vessel. The fire quickly spread through the oxygen-enriched air. By fire, flooding and suffocation, forty-two out of the sixty-nine crew died. The broken titanium hull, containing two nuclear reactors and at least two nuclear warheads, now sits a mile under the sea entombed in a concrete sarcophagus to prevent the toxic plutonium inside leaking.

US military intelligence first began to retrieve evidence of the Soviet's titanium-hulled submarines in the late 1960s. Satellite pictures of a submarine hull in Sudomekh Admiralty Shipyard in Leningrad (St Petersburg) revealed an unusual metal which seemed too reflective to be steel and which was not corroding. In the winter of 1969, Commander William Green, an assistant US naval attaché, was visiting Leningrad when he retrieved a piece of debris as it fell from a truck leaving the Sudomekh yard. It turned out to be titanium. Confirmation came in the mid-1970s when, while searching through scrap metal that had been sent to the US from the Soviet Union, intelligence officers found a piece of titanium inscribed with the number 705. This was known to be the serial number of the Soviet submarine project under surveillance. For a long time the US did not believe the intelligence they were receiving. Titanium seemed far too expensive and difficult to work with on the scale of the mammoth Soviet submarine hulls.

As the Cold War came to a close, the extreme deep-diving capabilities of titanium-hulled submarines were no longer necessary, nor were the Mach 3 speeds of titanium-framed aircraft. In the early 1990s, following the collapse of the Soviet Union, military spending on both sides of the Iron Curtain was cut; this was termed the 'peace dividend' by the US President George Bush and Britain's Prime Minister Margaret Thatcher. The last of the Project 705 submarines were decommissioned and funding for the Blackbird was removed, ending its life as the only military aeroplane never to be shot down or lose a single crew member to enemy fire.

Transitional titanium

Today titanium has a limited role. It is used on oil rigs and in refineries where the harsh maritime and chemical environments would quickly corrode steel; for limb implants where strength and biocompatibility are paramount; and for high-specification 'external limbs', such as bicycle frames, golf clubs and tennis rackets.[10] Titanium is still important in the aerospace industry, the prime consumer of the metal, where weight savings can significantly reduce fuel consumption.[11] But cheaper strong and lightweight aluminium alloys are now competing with it in all but the most specialist applications. Concorde, that symbol of civilian supersonic flight, was largely constructed from aluminium. Titanium's reign as a wonder metal was over; it had changed the world but, having done so, had been made superfluous.

But while steel skyscrapers rise out of the ground at an ever-increasing rate, we rarely see titanium used in the structures of modern society. One exception to this is the magnificent titanium-plated Guggenheim Museum in Bilbao, northern Spain. The futuristic ship-like curves of the exterior were originally chosen to be clad in stainless steel, but architect Frank Gehry was not happy with the appearance. It would be too bright in the sun and too dark in the shade. He considered zinc, lead and copper and, a few days before his proposal for the building was made public, was sent a promotional sampler of titanium. The new material's reflective properties gave it a velvety sheen in all light conditions. This was the metal Gehry wanted to use, but it was just too expensive.

One day, with titanium in mind, a member of the design team noticed a sudden drop in its price. Russia, the world's largest titanium manufacturer, had dumped large amounts of the metal on to the market. Within a week, Gehry bought all the titanium he needed before the price rose again. In 1997, the Guggenheim, covered in 33,000 titanium panels, opened to critical acclaim.

Titanium metal will always be in the background, but it will never outmatch iron for its unique cheapness, ubiquity and versatility. Holding back titanium's widespread use is the dated process by which titanium metal is extracted from its ore. The Kroll process, named after metallurgist William Kroll, that originally unleashed titanium's potential as a metal in the 1940s, is still the most widely used method of production today. The Kroll process is extremely energy-intensive and so very expensive.[12] As a result, titanium is an order of magnitude more expensive than steel and so, except for the most specialist applications, cheap steel is preferred over titanium. When weight is the chief concern, aluminium is usually chosen.

Titanium metal did not find the widespread application in society envisaged in the 1950s and production today is only about one ten-thousandth that of steel. This is all the more surprising considering that titanium is the fourth most abundant structural metal, after aluminium, iron and magnesium.

But titanium in its pure metallic form is only one half of its story. When titanium combines with oxygen atoms, with which it naturally bonds, it becomes titanium dioxide and that is so common in modern society that we rarely realise it is there.

Bright white titanium

As a Londoner every summer I go to Wimbledon where, before a match begins, I survey the immaculately mown lawns and meticulously drawn white lines of the tennis courts. The players come out of the grandstand, dressed head to toe in white, a tradition that stretches back to the first Lawn Tennis Championship in 1877. Whiteness was a symbol of wealth in the nineteenth century; today, thanks to titanium dioxide, both the courts' lines and the players' attire are a brighter shade of white.

We seldom pause to consider that white is everywhere in the world in which we live. In white-walled offices we wear white shirts and work on brilliant-white paper. We eat white foods and we use whitening toothpaste, because we think of whiteness as clean and pure. Adding white colouring to skimmed milk has been shown to make it more palatable.[13] In almost every application, whiteness in the products we buy comes from the harmless additive E171, a code name for titanium dioxide. Using titanium dioxide, murky greys and pale yellows are turned to pure white, making life agreeable for the modern consumer.

I first learnt of the use of titanium as a whitening agent when I was Chief Financial Officer of the Standard Oil Company (Ohio) in the late 1980s. Quebec Iron and Titanium (QIT) was a subsidiary, formed in 1948 shortly after the discovery of the world's largest deposit of ilmenite, a titanium iron oxide mineral, in the beautiful Lake Allard region of Quebec.[14]

After the iron had been separated out to make steel, we sold the titanium oxide slag to be used as white pigment.

On my yearly visits to QIT, from the air I could see the full scale of the ilmenite deposit, stretching out over an area the size of a hundred American Football fields, against the spectacular backdrop of Lakes Allard and Tio. The growth of the company over the last three decades had been relentless. In the 1950s titanium slag production grew from 2,000 to 230,000 tonnes and iron production from 2,700 to 170,000 tonnes. During the 1960s and 1970s a series of modernisation and expansion programmes was implemented as demand for titanium and steel products increased. Today QIT produces 1.5 million tonnes of titanium dioxide each year. But this causes barely a dent in the estimated global reserves of almost 700 million tonnes. Production can easily be stepped up to meet rising demand and so, unlike iron and oil, there has been little conflict over titanium reserves.

Like steel skyscrapers and silicon chips, manufactured whiteness is all around us, once a symbol of wealth but now a ubiquitous symbol of modern life. But why, of all the colours we could choose, are humans attracted to white? For the answer, we must go back to our understanding of light itself.

Why white?

In August 1665, Sir Isaac Newton drew the curtains of his study at Woolsthorpe Hall, Lincolnshire, save for a slit through which a sunbeam shone into the room. In the path of the beam he placed a glass prism that, as the light passed through, painted a spectrum on the opposite wall. With characteristic rigour he measured the dispersion of light across the room and from his results produced a revolutionary new theory of colour.[15]

For 2,000 years, since Aristotle wrote *De Coloribus*, it was believed that all colours were made from varying combinations of black and white, the polar opposites in our perception of colour. According to this theory, the colours of the rainbow were actually added to white light by a prism itself. To disprove this, Newton used an identical prism to show that the spectrum could be recombined into its original pure white state. Newton had demonstrated that colour was an intrinsic property of white light; he had 'unwoven the rainbow'.[16]

Newton divided the spectrum into seven colours: red, orange, yellow, green, blue, indigo, violet – the choice of seven to accord with the seven notes of the diatonic music scale and the seven heavenly spheres. 'But the most surprising and wonderful composition,' Newton wrote, 'was that of Whiteness ... 'Tis ever compounded.'[17] White light is the master of the rainbow; it is the basis from which all other colours emerge.

Sunlight, as opposed to the circular golden sun itself, is white light and so contains the full rainbow spectrum of colours. The sun emits these different colours of light in different proportions which, when combined, give the perception of white light.[18] This is no coincidence: our eyes have evolved over billions of years so that they are adapted to make sunlight the whitest and brightest source of light. We see objects as white when they reflect different colours of light by the same proportions as are emitted from the sun. The colour white is essentially an imitation of sunlight; we paint objects white so that they are bright and outstanding.

In contrast, gold is a reflection of the circular sun in the sky, whose white image is turned golden by the dispersion of light rays in our atmosphere.[19] We worship the sun, and so place a high value on gold. But the white light of the sun is so pervasive, like the white walls which surround us, that it goes almost unnoticed.

Since humans first moved into caves, we have sought to create a safe and hospitable environment in which our families and societies can live and develop. We have built barriers between us and the earth, rain and wind, separating ourselves from nature. In keeping walls white, we assert our control over the Earth's destructive forces, which constantly batter, wear down and soil human constructions. The white interiors of our houses and office blocks create a brilliant glow, competing with that of the sun.

Beyond white

At first sight, the prevalence of whiteness from titanium dioxide, while surprising and important, lacks the flair of supersonic aircraft and deep-diving submarines. But on closer inspection, we find that titanium's glow is as technologically sophisticated as the titanium frame of the Blackbird.

Titanium dioxide is not only incomparably bright; it is also clinically clean. Tiny nanoparticles of titanium dioxide absorb UV light from the sun. Coated in these particles, a wall disperses UV energy, which kills the bacteria that sit on the surface. A very thin layer on a window pane is transparent, but still absorbs UV light, which breaks up dirt on its surface. The coating also makes the glass surface repel water, so that when rain hits the surface droplets do not form and a thin sheet of water carries broken-down dirt away with it.[20] Most recently, the same principle has been applied to produce self-cleaning clothes.[21] By capturing light as well as reflecting it, titanium dioxide creates the immaculate environment in which we choose to live.

Titanium dioxide's extraordinary properties also show up in its interactions with electrons. Like silicon, titanium dioxide is a semiconductor and so can be used to carry an electric current in photovoltaic cells (which are used for solar power generation). While silicon both absorbs light and contains electrons to carry electric current, titanium is only sensitive to ultraviolet light. This makes titanium dioxide very useful in sunscreen since UV light causes sunburn, but less useful in photovoltaic cells. To capture as much light as possible from the sun, a light-sensitive dye is applied on top of the titanium dioxide nanoparticles.[22]

Technological innovations often come from the most unexpected places, and for titanium photovoltaic cells we have to look back to the development of silver halide photography at the end of the nineteenth century. Both silver halides and titanium dioxide are, by themselves, insensitive to most of the visible spectrum of light. Early photographic emulsions were only sensitive to bluer tones of light, and light at the red end of the spectrum was not picked up at all. In 1873, the German photographer Hermann Wilhelm Vogel discovered that certain dyes would improve the sensitivity of the photographic plates to different parts of the light spectrum. The dyes absorb photons from the sun causing electrons to be ejected. The energetic electrons then interact with nearby silver halide molecules. Vogel's experiments led to much more accurate black and white photographs and, eventually, colour photographs. Similar dyes are used in titanium dioxide photovoltaic cells today, but rather than the ejected electrons turning silver halide grains to silver, the electrons are transported by the titanium dioxide semiconductor to electrodes, generating an electric current.

These photovoltaic cells are the latest twist in titanium's fascinating relationship with sunlight. Along with white paints and artificial colourings, which mimic the light emitted from the sun, and sunscreen, which protects us from the sun's harmful rays, titanium dioxide can now harness the sun's energy to create electricity.

Titanium is one of three post-war 'wonder elements', along with uranium and silicon. Each serves to show how the world can be transformed by a new application of the Earth's elements. Uranium did so most dramatically when its extraordinary energy was unleashed on the people of Hiroshima, but titanium, too, has shaped the post-war era in both its military and civilian applications. The influence of the elements stretches back thousands of years to the earliest human communities. It is therefore remarkable to think of the impact that these three elements have had within my lifetime.

After the Second World War, we were tempted to think that, because new uses had been found for them, these were 'wonder elements' which would continue to change the world; we were wrong. After the bombing of Hiroshima, great excitement was generated around uranium and its future applications in everything from a cure for cancer to climate control. Today, uranium is viewed with dread and uncertainty; the bright nuclear future that was dreamt up in the late 1940s has never been realised. Titanium, once seen as a high-performance structural metal, came to be employed in roles as mundane as Russian cutlery sets, and is now dominantly used as a ubiquitous whitener, something that was not envisaged sixty years ago.[23] And silicon, which first appeared in the public eye in 1948 when the invention of the transistor was announced, was ignored by most people in the ensuing years. Silicon, though, has had the greatest impact of all the post-war 'wonder elements', enabling the development of smaller, faster and cheaper computers that have placed immense processing and communication powers into our hands, an extraordinary transformation for a material that for many millennia was ignored as useless, worthless sand.

SILICON

On a small stretch of beach to the south of Acre in modern-day Israel, the Na'aman River, heavy with silt, meets the Mediterranean. When the tide retreats, clean white sands, rich in silica, are revealed. Here, recounts Vannoccio Biringuccio, the medieval metallurgist, a group of merchants 'driven there by the fortunes of the sea' stopped in order to eat.[1] Finding no stones on the beach, the crew took lumps of nitre, a type of soda, from the ship's hold to support their cauldrons. 'In cooking the food they saw the rocks of that place converted into a flowing, lustrous material … Thus gave beginning to glass making.'[2]

By chance, these merchants had transformed simple grains of silica, consisting only of atoms of silicon and oxygen, into an object of beauty. From this initial discovery, came a profusion of glass innovations: glass beads in ancient Egypt; glass vases in the Near East; and glass mirrors in Venice.[3] 'Through many experiments and by addition or subtraction at will,' writes Biringuccio, the ancients and moderns have 'done so much that one can perhaps believe it would be difficult to go any farther with this art. For, as is evident, an infinite number of beautiful things are made from it.'[4]

Throughout its 5,000-year history, the versatility of glass has enabled the creation of startlingly original objects of beauty. The ability to keep transforming glass in this way is not a result of the tools used by the artisan, which have remained largely the same. Glassblowing, invented in the first century in the Near East and perhaps the most important innovation in the history of glassmaking, is still the technique by which most decorative glass is produced today. Rather, continued innovation in producing deco-rative glass is a result of the intrinsic malleability of the atomic arrange-ment of glass. When grains of sand are melted with soda, the resulting liquid is so viscous that, as it cools into a solid, atoms cannot move into position quickly enough to form a regular crystal structure.[5] Silicon and oxygen become frozen into a disordered structure that resembles a liquid, and so can be shaped into almost any form.[6] The random atomic structure of glass also leaves room for atoms of other elements to be accommodated. By adding small amounts of impurities, glass can be made with different optical properties: transparent, translucent, opaque and opalescent. And all of this can be done in a limitless variety of colours.

The transformation of sand into glass was merely one of silicon's great-est contributions to human progress. The second, invented many millennia later, comes from pure silicon crystals. The unusual electrical properties of these crystals enable their use in photovoltaic cells, which generate elec-tricity from sunlight, and in transistors, which form the processing core of computers. The transistor is arguably the most world-changing of all of silicon's uses, placing extraordinary computational and communication power in the hands of everyone.

Harnessing the optical and electrical properties of this extraordinary element, humanity has created objects of great beauty and powerful tech-nology that continue to amaze, delight and inspire us. But the story of silicon for most of its 5,000-year history is the story of glass. And the most important place in the history of glass is Venice where, during the Renaissance, decorative glasswork reached its zenith as an art form.

GLASS

On a sunny April morning in Venice, I walked past the window of an antique shop off the Campo Santo Stefano and spotted the tail end of an elaborate art-deco elephant. The shop is in a very narrow alley and so, among the crowds of tourists, I could not step back to take in the full dis-play. Returning to get closer to the window, I passed a series of wooden cubicles, presenting glass vases and goblets, before the black and turquoise elephant appeared.[7] The delicate twist of its trunk makes crystal clear the artistry of the glass workers on the nearby island Murano, where most Venetian glass is made. Elephants, with many protuberances which must be crafted and individually joined to the hulking body, are some of the most difficult glass objects to create. I have been collecting them for about five years. I first became fascinated by glass animals when, in Venice, I met a former director of the Musée du Louvre, whose extensive collection I greatly admired. Years later, I discovered that my partner collected sculptures of elephants, among the noblest creatures on Earth; they are highly intelligent and form deep social bonds. Combining my partner's interest in elephants with my love of Venetian glass, our collection had already begun.

Even before entering the shop I knew I was going to buy the elephant. From my body language that must also have been clear to the shop owner; I probably ended up paying too much. I now own more than a hundred glass elephants, and the herd is growing. This particular one was created in 1930 by Napoleone Martinuzzi, working in the studios of the manufacturer Venini. Under Martinuzzi's artistic direction, Venini became an innovator of new types and forms of glass. By introducing clean and elegant shapes and an intense new colour palette, he contributed to a revival of the Murano glass industry.[8]

Murano's glass industry first took off after 1291, when the Great Council ordered the glass furnaces in Venice to be moved to the island as they were causing many fires in the city.[9] The same soda and silica used by Egyptian and Islamic glassmakers was imported into Murano along trade routes with the East. These commercial links also provided a readily available export market, giving local glassmakers an edge over their competitors. In the island's glass workshops, artisans created objects of outstanding beauty using the same blowpipes, crimps and shears that are still employed on the island today.

At Murano, 'the best glasswork is made,' wrote Biringuccio. 'It is of greater beauty, more varied colouring, and more admirable skill than that of any other place.'[10] Georgius Agricola, a contemporary of Biringuccio, also valued the artistry of the Muranese artisans. During the Feast of the Ascension, he admired the diverse range of glass objects for sale: 'goblets, cups, ewers, flasks, dishes and plates, panes of glass, animals, trees, and ships'.[11] In Renaissance Venice, glass became an extraordinary art form.

Made from common sand, glass was not rare, but the rare skill needed to create beautiful glass objects meant that they were valued as highly as many precious minerals.

Among the most important inventions to come out of fifteenth-century Venice was *cristallo*, a type of glass 'as colourless and transparent as quartz'.[12] Impurities gave most glass an unpleasant yellow, grey or green hue, but using high-quality silica, made from crushed quartz river pebbles, purified soda and innovative furnace techniques, Venetian glassworkers created a crystal-clear and highly sought-after product.[13]

Glass warfare

Some of the most highly sought-after Venetian objects, often produced with *cristallo*, were mirrors. Glass, especially with a thin reflective coating applied to its surface, was a far superior way of creating a mirror than using a polished metal sheet.[14] However, uneven surfaces and poorly reflective silvering left a lot to be desired. 'One sees someone else there rather more than oneself,' one fifteenth-century observer remarked.[15] By improving the glass base and using a new material for silvering, glassmakers were able to produce the most 'divinely beautiful, pure and incorruptible' mirrors, 'even if the price is excessive'.[16] Perfectly transparent windows and mercury-coated glass mirrors were shipped out from Venice to wealthy households in Paris and London. In the seventeenth century, these extravagantly embellished luxuries were the height of fashion, bringing light into the aristocracy's grand chambers and reflecting the beauty of the inhabitants. The finest Venetian mirrors were sold for astronomical prices. One piece, framed in silver, was purchased for almost three times more than a painting by Raphael. *Cabinets aux miroirs*, rooms lined with mirrors, became the wonder of the day, both as optical phenomena and displays of wealth. Queen Anne of Austria and Catherine de' Medici both had them built.

Unable to produce glass of equal quality, the French were spending huge sums of money to import decorative glass objects from Venice. The trade balance became unsustainable and so, in 1664, French politician Jean-Baptiste Colbert asked Pierre de Bonzi, the French Ambassador to

Venice, to tempt Venetian glass artisans to Paris. Bonzi replied, 'whoever might suggest that they go to France would run the risk of being thrown into the sea'.[17] The Republic of Venice recognised the economic importance of its glass industry and did everything it could to keep hold of its monopoly, closely guarding its skills and trade secrets. Glassmakers were given enormous privileges, such as the right to carry small swords, and were given a prominent place during the procession of the Feast of the Ascension, all to entice them to stay in Venice. Even a noble who married the daughter of a glassmaker would not have to forfeit his rank, a powerful gesture in a status-ridden city. But the Venetians did not rely on prestige alone to protect their trade. Any glassmaker leaving Murano would face severe consequences. The Council of Ten, the Republic's secretive executive body, had ruled that 'if any worker or artist should transport his talents to a foreign country and if he does not obey the order to return, all of his closest relatives will be put in prison'. If that failed, 'an emissary will be sent to kill him and, after his death, his family will be set free'.[18]

Eventually, Bonzi did find three venal, risk-taking workers who were willing to leave Venice in return for a huge sum of money. In June 1665, the workers arrived in Paris and set up their workshops on the Rue de Reuilly. Twenty more Murano glassworkers soon followed, enticed by large salaries on offer at the newly established Royal Company of Glass and Mirrors at Saint-Gobain, near Paris. The Venetian authorities duly retaliated by threatening to gaol their families and seize their property, but since too many workers had already left, their threats became increasingly empty. Colbert even arranged for some of the workers' wives to join them in Paris. However, in January 1667 a Muranese worker in Paris came down with a fever and died. Three weeks later, another worker died after complaining of intense stomach pains. The word on the street was that they had been poisoned by the Venetian authorities. Being Venetian, they believed in conspiracy, and so the workers began to return to Murano. For a little while this halted progress at Saint-Gobain, but it was too late. Many of Venice's trade secrets had already been passed on and production of French glass and mirrors took off.

In 1682, Louis XIV used the output of the thriving French glass industry to construct the greatest *cabinet aux miroirs* ever built: the Hall of

Mirrors at Versailles. Seventeen huge mirrors filled the giant arches of the hall, each consisting of eighteen smaller square mirrors. 'The mirrors are false windows facing the real ones and expand this hall a million times over so it seems almost infinite,' wrote the *Mercure Galant* gazette.[19] The mirrors got cheaper and bigger and, in 1700, the Royal Company cast a piece of glass that was three metres long by one metre wide. In France, the glass industry soon rivalled, and then overtook, that of Venice, whose trade secrets were increasingly leaking into neighbouring European countries through espionage and betrayal. In 1680, the Venetian Ambassador lamented: 'I have tears in my eyes when I see how these many factories, which by an admirable gift that Providence, nature and hard work had granted particularly to us, have been so transported and sustained by the unpunished spitefulness of a few of our fellow citizens.'[20] At the end of the seventeenth century, the Venetian glass industry was on the verge of collapse. Muranese glasshouses failed to keep up with technological and stylistic innovations, such as lead crystal glass, created elsewhere in Europe.[21] The Council of Ten made the situation worse by putting in place restrictive regulations to protect the property rights of glassmakers and maintain the price of glass products. By doing so they eliminated the flow of new ideas into Venice and stifled innovation.

In Great Britain, renowned Parisian glassmaker Georges Bontemps helped the glass factory of the Chance Brothers in Birmingham to acquire French workers and their know-how.[22] He later moved there himself to work after fleeing the French Revolution, taking with him his unique expertise in sheet glassmaking.[23] But in spite of the influx of experts and ideas, there remained a barrier to the growth of the British industry. Since 1746 glass had been taxed heavily and only the rich could afford to buy it. Few homeowners could spare the money for windows, especially as the government taxed the windows as well as the glass from which they were made. This tax was 'daylight robbery'. The *Lancet* described it as 'an absurd impost on light' and 'one of the most cruel the Government could possibly inflict on the nation'[24] because it was causing so much ill health. In 1845 the tax was repealed and the glass industry began to grow rapidly. 'The manufacture of plate glass adds another to the thousand and one instances of the advantages of unrestricted and unfettered trade,' wrote Charles Dickens.[25]

With a sudden abundance of cheap glass, architects began to experiment with glass on an unprecedented scale.

The Crystal Palace

Thirty thousand eager spectators stood below the glass roof of the Crystal Palace in Hyde Park, London. 'Around them, amidst them, and over their heads was displayed all that is useful or beautiful in nature or in art,' wrote *The Times*.[26] 'Above them rose a glittering arch far more lofty and spacious than the vaults even of our noblest cathedrals.'[27] The crowds were awaiting the arrival of Queen Victoria to open the 1851 Great Exhibition of the Works of Industry of All Nations. As the Queen was driven by carriage through the gates of the park, the towering glass and iron structure of the Palace came into view in front of her and a frigate on the Serpentine fired a salute. Fortunately, fears that the blast of the guns would 'shiver the glass roof of the Palace and thousands of ladies will be cut into mincemeat' were not justified.[28]

Inside the cavernous glass structure, the most accomplished and magnificent inventions and craft from around the world were on display. Exotic silks, jewels and spices from the East stood alongside the latest scientific contraptions of the West. Alfred Krupp's new cast-iron guns sat next to the daguerreotype and Fox Talbot's calotype. Henry Bessemer had yet to turn his attention to the improvement of iron and steel manufacture, but he demonstrated a new vacuum table to aid the grinding and polishing of glass and a variety of inventions for improving the extraction of juice from sugar cane.[29] One hundred thousand exhibits were on show. And in the centre of the Palace was a magnificent eight-metre-high British crystal glass fountain.

Visitors paid a shilling, an entire day's wage for a labourer at the time. Many people must have thought it was worth it, since over a quarter of the population of Great Britain went. Travel was still a luxury and photography still in its infancy, but everyone could experience first-hand the wonders of the world on their own doorstep. The Great Exhibition was not just an amusement park. It was a trade show and marketing venue for inventions of the Industrial Revolution with the transparent Crystal Palace acting as a giant shop window for the competing products of nations from around the world. Prince Albert, Queen Victoria's consort, had been persuaded to hold the exhibition and become its Patron by Henry Cole, a civil servant. He had been overwhelmed by the scale and magnificence of the 1849 Exposition of the Second Republic in Paris. Cole decided that Great Britain would have to go one better, putting on not just a national but an international exhibition. That sort of display, he assured Prince Albert, would confirm Britain's position as the leading industrialised nation and affirm its continuing global eminence.

The Palace, designed by the architect and engineer Joseph Paxton, was an ideal venue for an exhibition. Sun streamed in from every angle, displaying each exhibit in natural light. At the time, glass was not considered a building material by many and Paxton was one of very few engineers experienced in its architectural use; he had already built the world's biggest greenhouse, 70 metres long, for the Duke of Devonshire at Chatsworth House in Derbyshire.[30] That was modest compared to his design for the 600-metre-long and 150-metre-wide Palace, built with 3,300 iron columns and 300,000 panes of glass.[31] The Chance Brothers won the contract to provide the glass, having previously been employed for the greenhouse at Chatsworth. They received a special commendation for bringing in new practices from Europe through Bontemps and others, being praised for 'the liberality, intelligence and spirit of the enterprise which they have manifested at great cost and risk'.[32] Glass had really become a global industry.[33]

The Crystal Palace made London '100 per cent glass-conscious'.[34] The world suddenly became aware of the importance of glass, not just for objects of beauty, but also as a utilitarian building material. However, each pane of glass in the Crystal Palace was still made individually by hand and blown by human breath. To make flat glass for windows, spheres or cylinders were first blown, cut open and then flattened.[35] This was expensive to do and often resulted in windows marked with defects. Larger sheets of glass were produced by casting molten glass on to an iron surface, but without a perfectly smooth bed on which to set the glass the surface would be rough. Grinding and polishing the sheet was a costly process in which about half the glass was lost.

The widespread use of glass as an architectural material would have to wait for an extraordinary innovation. In 1952, Alastair Pilkington invented the float glass process. Float glass is produced by dipping an iron bar into a pool of molten glass and pulling it out to form a continuous sheet. This sheet is drawn over a bath of smooth molten tin so that the glass also sets smoothly and does not need further working. Pilkington's process produced perfectly clear glass at a fraction of the cost of traditional processes.[36] Soon glass fronts became the norm for new city buildings. A recently built glass-fronted skyscraper in London pierces upwards to a height of 310 metres looking, appropriately, just like a shard of glass.[37]

Glass is now used in profusion, creating objects of both astonishing beauty and everyday utility. Mirrors and glass windows are no longer the preserve of the rich; the world about us has become coated in transparent and reflecting surfaces.

Glittering surfaces

In *The Music Lesson*, a photograph by Hiroshi Sugimoto, two wax figures stand beside a virginal. Above the female figure, who is playing the instrument, an ebony-framed mirror reflects her face against the backdrop of the studio's chequered marble floor. In the same reflection, the three black legs of the camera tripod can also be seen. The photograph is an approximate staging of a painting of the same name by Johannes Vermeer.[38] In that painting, although much harder to make out, a leg and a crossbar of the artist's easel are reflected in the mirror. Using a mirror, both Sugimoto and Vermeer make us aware of the artist, self-consciously placing himself in the room alongside his subjects. Mirrors provide an alternative perspective on the scene, expanding our vista and adding a level of reality. The image might even seem to come from behind the solid glass. In Lewis Carroll's *Through the Looking-Glass*, Alice amuses herself by pretending that the image seen in the looking-glass is another world. As she walks through the looking-glass, she leaves everything behind and enters a world of imagination. Jonathan Miller, author of *On Reflection*, explains: 'Apart from the straightforward sensuous allure of things which shine, shimmer, glint, glimmer, gleam or flash, reflections appeal to us, rather as shadows do, by paradoxically representing states of affairs whose actual existence is somewhere other than where they seem to be.'[39]

The earliest mirrors were objects of wonder and introspection. Plato was the first Western thinker to discuss them: 'You can take a mirror and turn it in all directions: from less than nothing you will have the sun and the stars in the sky, yourself and different animals, furniture and plants, and all the objects you just mentioned. Yes, all of these apparent objects, without any reality to them whatsoever.'[40] The mirror image, Plato believed, was deceptive and distorted. Narcissus confused the illusion of a mirror of water with reality and became enchanted by his own reflection. Unable to stop gazing at the beautiful being he believed he saw on the other side of the water, Narcissus died. On the other hand, Socrates suggested that if a youth viewed himself in a mirror he would understand himself better.

Today images in mirrors give even the most outwardly confident a sense of wellbeing. There are very few people who can walk past a mirror without glancing at their reflection. I wonder what life would be like without our reflected images. But glass and mirrors have done far more than just change how we perceive ourselves within society. They have also transformed our view of ourselves within the Universe.

'If they had seen what we see'

In the summer of 1609, Galileo Galilei, at the time living in Venice, heard news of a device for 'seeing faraway things as though nearby'.[41] The device, he was told, was made from a tube with a piece of curved glass placed at each end. Galileo was intrigued, but wary: curved glass was known to distort, and placing two pieces of glass together was assumed to create further distortion. A master craftsman, Galileo bought glass lenses from a spectacle maker and set about constructing his own telescope. By the end of the summer, he had created one which could magnify by eight times. He demonstrated it to the Venetian lawmakers 'to the infinite amazement of all'.[42] With his telescope, Galileo observed objects never before seen by human eyes. He mapped hundreds of new stars and even, to his surprise, viewed mountains on the moon and moons around Jupiter.

At the time, the Earth was widely believed to sit at the centre of the

Universe. Spherical planets and stars would rotate around this centre in crystalline spheres. The irregular features of the night sky, which Galileo observed, contradicted long-held beliefs about the perfect nature of these heavenly spheres. The more accurate measurements of planetary orbits made by the telescope also drove a stake through the heart of the Earth; it was no longer the centre of the Universe. Galileo's observations provided the evidence for the controversial new model of the Universe, communicated by Nicolaus Copernicus in 1543, in which the Sun was placed at the centre.[43] That model not only contradicted prevailing religious doctrine, but was also an affront to common sense. How could the Earth move through space without its motion being felt by people on the ground?[44] The evidence provided by Galileo's new instrument, however, continued to stand up. For the first time since antiquity, the extent of the night sky was expanding. And with the knowledge conveyed by silicon-based telescopes, Western thought began to break away from a view of the Universe that had changed little since the time of Aristotle. 'If they had seen what we see,' Galileo wrote of earlier astronomers, 'they would have judged as we judge.'[45]

Silicon opened the heavens beyond the natural limitations of the eye. The further we saw, the further we wanted to see. Galileo's invention started a drive to build ever more powerful telescopes. By the middle of the seventeenth century, wealthy astronomers were building telescopes up to 50 metres in length, requiring a system of masts and pulleys for operation. The greater length of telescopes was one way of overcoming the blurring of the image which resulted from the curvature of the lens.[46] In these 'refracting telescopes', different colours of light were bent by a different amount as they passed through the lens, producing an unclear image. Isaac Newton overcame this problem by inventing a 'reflecting telescope' that used mirrors rather than lenses. The mirrors reflected each part of the incoming light in the same way, regardless of its colour, so that a clear image was produced. To Newton, his telescope was further proof that white light is composed of a spectrum of colours.

Even when very large mirrors were used, the image produced remained clear. And the bigger the mirror, the further the telescope could see. Eighteenth-century astronomer William Herschel took this principle to extreme lengths.[47] He did more than any astronomer to improve the power of reflecting telescopes, extending our view of the Universe far outside the Solar System. 'The great end in view,' he wrote to Sir Joseph Banks, President of the Royal Society, 'is to increase what I have called "The Power of Extending into Space ..."'[48] By grinding and polishing larger and larger mirrors, Herschel was eventually able to resolve starry pinpricks of light into diffuse objects.[49] Some of these 'nebulae' were later shown to be galaxies akin to the Milky Way. Reflecting telescopes have grown in size ever since: from the two-and-a-half-metre circumference mirror in the Hooker telescope on Mount Wilson in 1917, to the five-metre Hale telescope on Palomar Mountain in 1948. Today, telescopes, with mirrors of more than 10 metres, sit high on mountains in the Canaries and Hawaiian islands, recording the night sky around us with unprecedented accuracy.

Photons not only carry information about the stars from which they originated, they also carry energy. Long before the invention of the telescope, mirrors were used to capture and concentrate the light energy emitted from our own star, the Sun.

SOLAR POWER

In the middle of the seventeenth century, Father Athanasius Kircher, a Jesuit scholar, positioned five mirrors so as to direct sunlight on to a target 30 metres away. The heat produced was so intense that his assistant could not comfortably stand at the target. 'What terrible phenomena might be produced,' Kircher wondered, 'if a thousand mirrors were so employed!'[50]

Kircher would probably have been familiar with the legend of Archimedes' burning mirrors. At the start of the third century BC, as the Roman ships of General Marcellus advanced towards Syracuse, Archimedes directed his soldiers to raise and tilt their reflective shields towards the armada. The result was dramatic; the concentration of heat was so intense that the ships were set alight. Burning mirrors were among a large armoury of imaginative inventions that Archimedes deployed to defend Syracuse against the Romans. With his knowledge of geometry, Archimedes could calculate how to focus light rays and also aim projectiles to destroy the enemy's ships before they could get close enough to land to do damage.[51]

In his sixteenth-century book *Pirotechnia*, Vannoccio Biringuccio recalls a conversation with a friend who had created a mirror almost 70 centimetres across. One day, while watching an army review in the German city of Ulm, the man entertained himself by using his mirror to direct sunlight on to the shoulder armour of one soldier, creating so much heat that 'it became almost unbearable to the soldier ... so that it kindled his jacket underneath and burned it for him, cooking his flesh to his very great torment'.[52]

In the sixteenth century, Leonardo da Vinci designed some novel peacetime applications for the sun's rays. Ambitious as always, Leonardo planned to build a six-kilometre-wide concave mirror that would focus sunlight on to a central pole to heat water or melt metals.[53] As with so many of his inventions, this particular monstrous device never made it beyond his sketchbook. It was not until the Industrial Revolution in Great Britain that glass and mirror constructions were possible on a bigger, but not quite Leonardo, scale. In his later years, Henry Bessemer built a solar furnace for the smelting of metals. Inside a 10-metre-high tower, a reflector directed sunlight on to a four-square-metre concave mirror in the roof. This focused the light through a lens at the bottom of a tower and into a crucible. He managed to melt copper and vaporise zinc in this furnace, but it was not very efficient and cost a great deal to build. After some years, even Bessemer 'became disheartened, and abandoned the solar furnace'.[54]

Across the Atlantic in Philadelphia, an American inventor, Frank Shuman, turned his attention to the problem of concentrating solar power. At the turn of the twentieth century, using the heat-trapping properties of glass, he raised the temperature of water in something he called his solar hot box to just below boiling point, even when there was snow on the ground.[55] 'I am sure it will be an entire success in all dry tropical countries,' he wrote. 'It would be a success here on any sunshiny day; but you know how the weather has been.'[56] In Egypt, where the weather was rather more reliable, his solar hot boxes powered steam engines used to pump water for irrigation. Another inventor, Aubrey Eneas, built a series of giant cone reflectors, some over 65 square metres in area, to collect solar radiation in the intensely sunny states of California and Arizona. Eneas was also inspired by the parabolic trough reflectors invented by John Ericsson, the Swedish-American engineer who built the *Monitor* ironclad during the American Civil War, but who also devoted the last twenty years of his life to building solar machines. Both Bessemer and Ericsson, pioneers in the production and use of iron, were concerned that the coal supplies they used to smelt iron ore and power steam engines would run out, and so they sought alternative sources of energy. Eneas's plan was to provide a cheap energy source for those living in the desert, far from traditional coal supplies. By increasing the scale of their systems, both inventors had hoped to produce cheaper solar power; but even at scale, these concentrated solar power systems could not provide electricity which was competitive with that generated from conventional sources. And that remains the case today. On the arid plains near Fuentes de Andalucía, Spain, there are more than 2,500 mirrors, each with an area of 120 square metres, directing sunlight towards a tower placed at their centre. In the tower, molten salt is heated to almost 600 degrees centigrade. The molten salt can be stored in tanks until it is needed, when it can be used to drive steam turbines and generate electricity. But without very large subsidies, even this modern plant is uncompetitive.

Solar power was forgotten until shortly after the Second World War, when scientists at Bell Laboratories in New York began to investigate some unusual electrical properties of silicon. The research of Gerald Pearson, Daryl Chapin and Calvin Fuller led, in 1954, to the creation of the first silicon photovoltaic cell.

Photovoltaics

Daryl Chapin had been tasked by Bell Laboratories with developing a new portable power source which would power their telephone systems in tropical climates, where traditional dry-cell batteries degraded quickly. He began to investigate wind machines, steam engines and solar energy as possibilities. Rather than trying to capture the Sun's energy using mirrors and heat boxes, Chapin decided to investigate another medium for harnessing solar energy, known as the photovoltaic effect.

Alexandre-Edmond Becquerel, the father of Henri Becquerel (of radiation fame) discovered the photovoltaic effect in 1839.[57] Becquerel placed two brass plates in a conductive liquid and shone a light on them. He noticed that the light caused an electric current to flow in the solution. If this current could be used, the energy of the Sun could be harnessed.

Over a hundred years later, scientists had still only succeeded in harnessing one two-hundredths of incoming sunlight using photovoltaic cells. This did not make enough power for Chapin's needs and so he began to search for alternatives. Word of Chapin's work reached Gerald Pearson and Calvin Fuller, two other scientists working at Bell Labs, who were experimenting with the unusual electrical properties of silicon semiconductors. They thought the materials they had been developing could be used to create a photovoltaic cell. To their surprise, their idea not only worked but also created a photovoltaic cell that was five times better than anything else available.[58]

In April 1954, they announced the invention of the Bell Solar Battery, demonstrating to journalists how it would be used to power a radio transmitter. It quickly began to prove its value in providing energy to Bell's developing markets in the tropics. Solar cells got their first big break, though, when they were used in the American Vanguard space programme in 1958. While the vehicle's chemical batteries were rapidly depleted, the solar unit continued to function years after the launch. In satellites, solar cells had found their first major market.[59]

Even today, solar cells are often the most cost-effective way of generating energy for remote regions, avoiding the costly infrastructure of setting up power lines and transporting fuel. Their versatility allows them to be installed as single units in remote, energy-poor regions. In 2001, I visited Indonesia to see BP's solar rural electrification project, at the time the fifth largest of its kind in the world. Small-scale silicon solar cell arrays were used to generate electricity for almost 40,000 homes in village communities. Electric water pumps can be used to irrigate crops and electric lighting has been brought into homes, schools and medical centres. Solar cells have also indirectly improved education and learning. As I saw, children could study not only during the day but also at night.

Unlike fossil fuels, which occur in pockets dotted about the Earth, the Sun shines everywhere. In one year, more energy reaches the Earth's surface from the Sun than will ever be extracted from all sources of coal, oil, natural gas and uranium. In one day, the Earth's surface receives 130,000 times the total world demand for electricity. Despite this, solar energy still accounts for only a thousandth of total global electricity production. Part of the reason is that harnessing the energy of the Sun is notoriously inefficient. A small electric current is produced each time a photon of light is absorbed by a silicon solar cell. This is because the photon's energy is transferred to an electron and its positive counterpart, called a 'hole', in the cell.[60] However, while all the energy of a photon is transferred when it is absorbed, not many photons are absorbed in the first place. The photon must have just the right amount of energy for this to happen, and only a fraction of photons do. As a result, even the very best laboratory-based solar cells capture and convert only 40 per cent of the light that falls on them into electricity. The cells used in normal commercial applications convert between 10 and 20 per cent. That still makes them several times more efficient than the first solar cells, created at Bell Labs in 1954. That improvement, in the space of only sixty years, is remarkable; after billions of years of evolution, plants, which convert light into stored energy through photosynthesis, have only developed an efficiency of 3 per cent.

The greatest barrier to the success of the solar cell, however, has not been technical, but economic: solar cells have produced costly electricity because they have been expensive to manufacture. This is beginning to change as new technologies, such as using cheaper offcuts from the silicon used in chip manufacture, begin to mature. The cost of manufacturing, too, has fallen rapidly, in large part because of the economies of scale obtained by Chinese producers in China's growing market. Nonetheless, electricity generated from solar cells has not yet reached grid parity, the point at which cells would be economically competitive with traditional non-renewable fuel sources. That point, though, is getting closer. As more solar cells are made they generally become cheaper; in 2011, manufacturing capacity increased by almost 75 per cent on top of an average annual growth rate of 45 per cent over the past decade. This continued growth will be vital in the transition to a low-carbon energy economy.[61]

When they announced the invention of the silicon solar cell by Bell Labs in 1954, the *New York Times* wrote that it marked 'the beginning of a new era, leading eventually to the realization of one of mankind's most cherished dreams, the harnessing of the almost limitless energy of the sun for the uses of civilization'.[62] That dream may become a reality, and it will be one free of emissions of greenhouse gases. There is still a long way to go before the scale of solar energy comes close to that of fossil fuels or nuclear energy, but, of all renewable energy sources, solar has so far proved itself to be the most promising.

COMPUTERS

Anchorage, Alaska, 1970: red lights were flashing wildly on the control panel. The core memory of the computer had just crashed. Back then a computer crash was literally a crash, the spinning mechanical disks grinding together to a halt. The constant restarts made running even the simplest program extremely arduous. It would be a long night. I was working as a petroleum engineer in my first real job. Because of my experiences at Cambridge University I was, in those days, one of a handful of people who knew how to make the solution of engineering problems easier by using a computer. I was working to an extreme deadline. My boss was about to go to a meeting with some very powerful men from some even more powerful US oil companies. They were to discuss how much of the giant Prudhoe Bay field each of the participating companies owned. He wanted me to find an answer and to make sure that he, from a then small company, impressed the other bigger companies with his technical prowess.

It was not a straightforward problem. The companies owned leases of different pieces of land on top of the field and so what each one owned depended critically on the areal distribution of the oil, and that turned out to be very uneven. A long night turned into an early breakfast when, after a lot of stops and starts, a solution appeared and I went to the office. Overnight working turns out not to be a new phenomenon.

I was doing this in Anchorage's only 'computer bureau' run by Millet Keller, a graduate of Stanford University. It contained only a single computer, an IBM 1130, the state of the art at the time. During the day, Millet would run commercial programs in computer language COBOL, creating financial accounts for local banks. At night, I was able to run my own programs written in FORTRAN, a popular scientific and engineering computer language. Using the advanced IBM technology, which performed 'as many as 120,000 additions in a second', I was able to model BP's Alaskan oilfields to help in their development.[63] BP has been an innovator in computer technology since its earliest days. In the early twentieth century, it developed programs to calculate the most efficient routes for oil tankers to travel. But the invention of the IBM 1130 provided a huge leap in the processing power available to the industry. As a geophysicist, Millet was interested in the work I was doing and would often stay up to work with me, watching over the temperamental machine or feeding in the next punch card.

The IBM 1130 was the first computer I had encountered outside Cambridge University. It was less powerful than Titan, the university's monstrous mainframe computer, and far smaller, cheaper and more accessible. Titan had filled an entire room and required a whole laboratory team to work it. IBM sought to take computing technology out of that sort of setting to a wide variety of industries for which computing was becoming a necessity.

Today, exploring and drilling for oil without computers is unimaginable. By the time BP was producing oil from the Thunder Horse field, whose namesake production platform nearly sank as Hurricane Dennis passed close by in 2005, it had been able to use seismic and other data to construct a three-dimensional visualisation of the reservoir, several kilometres below the surface.[64] That allowed teams of scientists and engineers to work together and actually see the implications of a decision, such as where to place a well, on the long-term health of the reservoir. Processing the huge quantities of data needed to do this has been made possible by the extraordinary growth of a computer's capability over the last sixty years. And at the core of all this technology, back in Anchorage in 1971 as in the high-performance computer age of today, is a simple, tiny device made from silicon: the transistor.

Silicon transistor

In the late 1940s, William Shockley and his team in the solid state physics group at Bell Labs were exploring the unusual electrical properties of a group of elements called semiconductors. Bell's telephone networks were still operated using mechanical switches and signals were amplified using vacuum tubes.[65] These were slow and unreliable and so the director of research was tasked with finding an electronic alternative. Shockley thought the answer could be found in semiconductors, from which he hoped to create an amplifying and switching device.[66] Although its theoretical basis seemed flawless, it did not work. His colleague John Bardeen, a brilliant theoretical physicist, then set his mind to the problem. He realised that electrons were becoming trapped at the surface of the semiconductor, stopping current flowing through the device.[67] Working with Walter Brattain, whose skilled hands matched and complemented Bardeen's brain, Bardeen was able to overcome the surface trapping and, in doing so, turned Shockley's idea into a practical reality, the world's first transistor.[68]

At the end of June 1948, Bell Labs announced the invention of the transistor by Shockley, Bardeen and Brattain; they would later win the Nobel Prize in Physics for this breakthrough. At the press conference, they explained that the transistor had the potential to replace the vacuum tube, the device then used to make radios and rudimentary computers. Like the vacuum tube, the transistor could amplify electrical signals and act as an on-off switch, but do so much faster, in a much smaller volume, using much less power.[69] At the time, the media thought all this unimportant and made little fuss. The *New York Times* 'carried the big news on page 46, buried at the end of a column of radio chitchat'.[70] The potential for the transistor to change the world had yet to be realised by the wider public. After all, journalists must have wondered, what impact could these devices and their abstract functions have on our everyday lives? Even today, few people make a connection between these minute pieces of silicon and the complex functioning of the computers, with which we create images, manage communications and generate sounds.

Any computational problem can be broken down into a set of simple logical steps, such as the decision to combine two numbers or choose one or the other. These steps are controlled by 'logic gates', which are the basic building blocks of digital circuits. Logic gates are made from transistors and other simple components, and they use transistors as switches to send signals. Most logic gates have two on-off switches which together act as inputs. Each switch can either be off or on, known as '0' or '1', and the logic gate's output is determined by these two inputs as well as the type of logic gate. For example, an 'AND' gate will give an output of 1 only if both the first 'and' second inputs are 1. All other input combinations (0 and 1; 1 and 0; 0 and 0) will result in an output of 0. A computer, at a fundamental level, is simply a number of these transistor-based logic gates linked together to produce a complex output. The capability and complexity of the computer rises as more and more gates are connected.

Transistors allow this to happen because they are very small, very cheap and use only a little power. Those features allow enormous numbers of them to be put together in one computer. It is, however, their speed that makes computers really useful. A transistor's on-off switching function is controlled by a small electric current. Its small size and the speed of the electrons enable it to be turned on and off well over 100 billion times each second. If you used your finger, it would take around 2,000 years to turn a light switch on and off as many times. Silicon's semiconducting properties make it ideal for making these switches, although other semiconductors, such as germanium, were originally used and today transistors can be made of many different alloys. None of them, however, rival silicon's combination of high performance and low cost.[71]

The first commercial applications of the transistor were not, however, computers, but technologies that used its other function as an amplifier. The first of these was a hearing aid produced by Sonotone in 1952. The same principle was applied in radios, amplifying the electromagnetic waves received from transmitting stations. The small size of the transistor dramatically reduced the size and cost of radios, making them portable and opening up their ownership to a vast new market. The transistor radio, the 'trannie', heralded a new era of popular music heard anywhere by everyone. As these new products took off, the importance of the transistor began to be widely recognised. In March 1953, *Fortune* published an article entitled 'The Year of the Transistor'. 'In the transistor and the new solid-state electronics,' *Fortune* wrote, 'man may hope to find a brain to match atomic energy's muscle.'[72] Silicon had joined uranium and titanium in a class of post-war 'wonder elements'.

Silicon chip

Soon after Shockley, Bardeen and Brattain had invented the transistor, their relationship began to break down. Shockley, paranoid and competitive in the extreme, felt he was not being given sufficient credit for the invention.[73] He became unhappy at Bell Labs where his apparent lack of management abilities, combined with his foul temperament, led him to be overlooked for promotion. In 1956, he left to set up his own company, Shockley Semiconductor, in California, having been encouraged to move there by Frederick Terman, the Dean of Stanford's School of Engineering. Terman had the vision to see the potential of the semiconductor industry and wanted his graduate students to become a part of it. Together, Shockley and Terman shifted the centre of the industry from the East to the West Coast of the US, laying the foundations of Silicon Valley.

At Shockley Semiconductor, a number of brilliant individuals began investigating the potential of silicon. 'Neither the processing nor the physics of [silicon] was well understood,' wrote Gordon Moore, an employee at the company. 'We were just exploring the technology and figuring out what could be done and we had a lot of things to make work before we could try and build something.'[74] However, working at Shockley Semiconductor was a trying experience. Shockley's bad temper and poor management practices wore down his workforce. He was known to stage public firings and demand lie-detector tests over trivial matters. Shockley and his employees disagreed not only about the direction of the company but also about which new inventions should be commercialised. After about a year, a group of eight of Shockley's most talented and ambitious employees decided to leave the company. The 'traitorous eight' were put in touch with Bud Coyle and Arthur Rock, the latter the father of venture capital. Coyle and Rock persuaded them that, rather than being employed by another company, they should set up on their own. With US $1.4 million of funding from Sherman Fairchild, an inventor and businessman with a large stockholding in IBM, the group founded Fairchild Semiconductor, in nearby Palo Alto, California.

At the time, one of the biggest technical problems was making the use of a transistor very reliable, unlike the cumbersome and unreliable vacuum tube. Each transistor had to be connected in a circuit by wires installed by hand. As the number of circuits in a computer grew the chances of one of these connections failing rose significantly. That was very risky. Other components in the circuit, such as resistors, were not made of silicon but of carbon and other materials. This made circuit production an expensive and inefficient process.[75] In 1958, Jack Kilby, a scientist at Texas Instruments, started making changes to these circuits that would eventually lead to the development of the 'integrated circuit' by making all the components out of silicon. But Kilby's circuits were still connected with fine wires.

At Fairchild Semiconductor, a method had recently been developed to package and to protect these components by using the silicon dioxide layer that naturally forms on the surface of silicon.[76] Robert Noyce, a co-founder of Fairchild, described the process as being like 'building a transistor inside a cocoon of silicon dioxide so that it never gets contaminated. It's like setting up your jungle operating room. You put the patient inside a plastic bag and operate inside of that, and you don't have all the flies of the jungle sitting on the wound.'[77] Noyce began to think about what else could be done with the new process. He realised that the oxide layer could be used to simplify the production, and so could reduce the cost, of entire electronic circuits. The insulating properties of the layer enabled all the components of a circuit to be produced simultaneously on one piece of silicon. Instead of using wires, the components of the circuits could be connected by using a thin sheet of metal spread on top of the dioxide layer. Wherever a hole was punched through it an electrical connection would be made with the component underneath. Electrical connections could now be 'printed' on to a circuit, rather than made with fragile wires. He called this invention the 'integrated circuit', in which transistors, capacitors and resistors were printed and connected simultaneously on a piece of silicon.[78] This dramatically increased reliability, so much so that NASA used integrated circuits on the early Apollo space missions. Production costs were also dramatically reduced. Bardeen believed that recognising the natural tendency of silicon to form a protective dioxide layer had led to an invention which was as 'important as the wheel'.[79]

Moore's law

Fairchild Semiconductor became the leader in the development and production of integrated circuits as a result of Noyce's invention. The company grew rapidly; revenue was US $500,000 in 1958 but was over forty times that by 1960.[80] Around it many more computer technology and software companies were created and their location came to be known as Silicon Valley.

In 1965, Gordon Moore, one of the Fairchild 'traitorous eight', noticed a consistent trend in the way in which the price and size of silicon transistors fell; that trend has underpinned the extraordinary growth of Silicon Valley ever since. Moore's eponymous law states that the numbers of electronic devices (such as transistors, resistors and capacitors) which can fit on to a computer chip will double every year.[81] In 1965, Moore expected that this rate of increase would continue for at least ten years so that by 1975 the number of components that could fit on to a computer chip would have grown from 60 to 60,000. To everyone's surprise he was right. But the realisation of Moore's law did not stop in 1975. The exponential rate of increase in computing power and the consequential reduction in the cost of that power has been going on ever since.[82]

Today the most advanced microprocessors contain over two and a half billion transistors, the size of which has now shrunk to an incredibly small 22 nanometres, just nine times wider than a DNA chain. When combined with the overall growth of the computer industry, this leads to an extraordinary result: more than 10^{18} (a one with eighteen zeros following it) transistors were produced in 2011. This is more than the number of grains of rice grown across the world each year and more than the world's yearly output of printed characters. It costs even less to produce a transistor than it does to print each letter in a book, newspaper or magazine. The process of miniaturisation, described by Moore's law, produces faster and cheaper chips. And when chips became smaller and cheaper, they were used in more and more devices and were embedded into our daily lives. As Moore wrote in the article in which he first outlined his law: 'the future of integrated circuits is the future of electronics itself'.[83]

In 1968, Moore and Noyce were bought out of Fairchild and used the money to create their own company: Intel. I joined the board of Intel in 1997 on the suggestion of Mike Spence, the Dean of Stanford's Graduate School of Business. I had been chairman of the school's advisory board, having studied there. However, I wanted to stay involved in California's thriving business sector and believed I could learn a lot at Intel. Before I joined the board, I met Andy Grove, Intel's CEO, who had worked with Noyce and Moore at Fairchild. Grove remains one of the most impressive business thinkers I have ever met. He had the intellect and dynamism to execute successful strategic plans, again and again, in the fast-paced semiconductor industry. But Grove also deeply understood the science behind Intel's products; he had written textbooks on semiconductor physics. With the veteran venture capitalist Arthur Rock as a member and the master engineer Gordon Moore as its chairman, the board was formidable; the company's management was world class. Grove's mantra was 'only the paranoid survive'.[84] He acted that way and he made the board and management follow his lead. In this fast-moving industry you had to be aware of what changes were on the horizon. More than that, you had to check that you were on top or ahead of them. Grove called the most important of these changes '10x forces' because the change 'becomes an order of magnitude larger than what that business is accustomed to'.[85] The invention of the integrated circuit brought about one such change. More recently, the invention of the internet has brought about another.

Silicon communication

In the early 1990s, at the European Centre for Nuclear Research (CERN) near Geneva, Tim Berners-Lee, a computer scientist, was trying to find a way to help CERN's thousands of scientists work together more effectively. Each of their experiments on particle collisions created vast quantities of data, but without a communication network to share information it was impossible to collaborate substantively. A lot of research had already been done on the theory and practical design of information sharing networks by the US military. In the 1950s, tensions in the Cold War called for a decentralised communication network. If communication relied on a single line running from one point to another it would be capable of fatal interruption. However, if that line formed part of a much larger interconnected network, messages between points could be easily re-routed along different paths, thereby providing diversified back-up.

Building on this work, Berners-Lee created a system to link the computers at CERN; this later became the World Wide Web. Those in academia and industry were the first to use it to link the work of many collaborators. They already recognised the importance of computers; the immense processing power of these silicon-based machines was used in complex tasks such as mapping oilfields and simulating climate models. But Berners-Lee's invention also made possible a simple way of establishing an easy-to-use global communication network, whose use soon spread to the general public. That was the internet whose birth coincided with the rapid increase in the number of people owning a personal computer. By July 1995, 6.6 million computers were connected; the following year that number had almost doubled. Shortly after I joined Intel's board in 1997, Andy Grove announced that Intel should lead the way in the creation of a billion connected personal computers around the globe; at the time it was difficult to believe that it would happen.

Today, the internet connects over two billion people, fulfilling and surpassing Grove's vision. And the network even extends out into space, connecting astronauts on board the International Space Station to the Earth. The creation of the internet also added a new dimension to silicon, initially used just for computers but now also for communication infrastructure and devices. Berners-Lee's innovation relied upon the silicon infrastructure laid down by Shockley, Moore and the other Silicon Valley entrepreneurs. It also depended on silicon in a very different format. Silicon optical fibres, a type of glass, were developed in the 1970s and 1980s, replacing copper for long-distance telecommunication lines. As a result, the capacity of these lines was expanded hugely, which enabled the internet to transmit information across the world at the speed of light.[86]

As the backbone, the internet was a necessary means of meeting the increasing demand by people in business and everyday life to communicate with each other in real time. However, something else was needed to satisfy this demand fully: the interface between the human and the machine which needed to be made simpler and more pleasurable to use. This very need was recognised by Apple and that has made its name over the last couple of decades. In May 2012, I met Sir Jony Ive, senior vice-president of industrial design, in the Silicon Valley town of Cupertino, the home of Apple.

Form and function

I met Jony Ive outside, in a sunlit courtyard. We sat down for coffee and he began to elaborate on the process of design. 'At its core, my job is to think about the relationship between function and form,' he explained, before pointing to the table in front of us: 'Look at this cup. When we drink from it, we don't think about it because we know exactly what to do with it. Its form is intrinsically tied to its function. But around the time of the Industrial Revolution, this began to change. Mechanised objects created a disjoint between function and form, to the point where, in the smartphone today, there are an extraordinarily large number of functions with no intrinsically associated form.'

A smartphone is powered by electrons zipping across atomic layers of silicon, but its complexity is hidden away behind a shiny metal case and sleek graphical interfaces. This exterior is as critical as the technology inside; it ensures smooth and flawless functioning that makes our interactions with computers as easy as those with a teacup. The first personal computers looked threatening and scared off many potential users. They looked like laboratory equipment, beige and black boxes, designed by scientists for scientists. That created a barrier between the user and the computer, a barrier that Jony has spent his career at Apple breaking down. Later in the day, he took me into the entrance, but no further, of his design studio, where he directs the small team that creates the designs of the future. Frosted glass windows keep prying eyes out of the hushed and calm environment. Here ideas are plenty, but masterpieces, his standard, are few. Days, weeks, even months, are spent designing, modelling and re-forming each and every button and curve to create objects that transcend utility; they are attempting to create objects of desire. That desire and power have spread throughout the world, with truly revolutionary consequences.

Social media revolutions

In the middle of December 2010, Mohamed Bouazizi was selling fruit from a street stall in Sidi Bouzid, Tunisia, when he was confronted by two police officers. He did not have a licence, or the money to pay the usual bribe expected by the officers, and so his cart was confiscated. He tried to lodge a complaint at the local governor's office, but they just laughed at him. Helpless and in despair, he returned with a can of petrol, doused himself with its contents and lit a match. The news of his death spread quickly and triggered a series of protests, many of which were organised through the use of social networking sites. The president fled the country shortly after.

Twitter and Facebook had provided a platform through which Tunisia's despairing youth could communicate their shared grievances and coordinate political action. Around the Arab world, similar extraordinary protests followed, with the eventual demise of rulers in Egypt, Libya and Yemen. This was not the first era in which silicon has been used as an enabler of political revolution. Transistor radios were used during the Cold War by Radio Free Europe to broadcast anti-communist propaganda across the Soviet Union. A revolutionary group always tries to take control of state radio stations in an attempted *coup d'état* as, for example, Hugo Chávez did in the early 1990s. Whoever controlled these controlled the country.

Mobile communications and the internet, with their ability to spread information very widely, allowed the revolutions of 2011 to build momentum far more quickly and with greater effect. Silicon enabled this to happen by providing the tools to have debates and discussions. That is a noble purpose, but these tools also enable surveillance, snooping and persecution. In Tunisia authorities used the internet to target and arrest prominent bloggers before the revolution. In China and Russia much of the information accessible online is censored and those using social media to voice opposition to the government are kept under surveillance, usually for a bad purpose. Just as with other of the other elements considered here, silicon can make the good and the bad happen. And it can do so very rapidly without geographical boundaries.

Silicon society

'What news on the Rialto?' asks Shylock in Shakespeare's *The Merchant of Venice*. During the Renaissance, the Rialto was the financial and commercial centre of Venice, and to find out what was going on there you had actually to go and see for yourself. Local communication was limited to a walking pace, while international communication occurred only as fast as a ship could sail. Centuries later, during the Industrial Revolution, humanity harnessed the energy in coal and oil to power steam trains and ships, and then cars and aeroplanes. By transporting people further and faster, carbon expanded our geographical horizons and our capacity for communication.

But it is silicon that has enabled the most recent and dramatic step change in our power to communicate. Just as with carbon-based transportation, silicon has transformed our individual daily lives, giving us greater choice over who our 'friends' are and enabling us to keep in touch with a much wider social network. The power of silicon, though, extends far beyond that of carbon. Even today around only 15 per cent of the world population has a car; far fewer have ever flown by aeroplane.

Silicon is far more pervasive in its use in mobile phones. These have dramatically changed the way societies are developing. They have existed for a long time but, just as with the early computers, they were expensive, bulky and used a lot of power; the batteries were so large that they had to be placed in the boot of a car. Individuals can now access computing power that was once only available to universities and big business: a smartphone contains more computing power than that at the whole of NASA when man landed on the Moon in 1969. By the 1990s, mobile phones had become so cheap that they were affordable to many in the developing world. By 2002, there were over one billion mobile phone subscriptions, a milestone which took fixed telephone lines 128 years to reach. Today, around 75 per cent of the world has access to a mobile phone. By connecting so many previously isolated voices, silicon has shifted the balance of power within society. One only needs to look to the growing influence of NGOs and internet-based lobby groups to see how political change in nations across the world is being catalysed by silicon.

To find out the important news of the day, you no longer have to go to the Rialto; you just simply reach into your pocket. Silicon has enhanced our ability to understand the world, doing for the human brain what carbon and iron did for human muscle.

The futurologist Ray Kurzweil points to 'a future period during which the pace of technological change will be so rapid, its impact so deep, that human life will be irreversibly transformed'.[87] At this point, humans, machines, physical reality and virtual reality will be melded together. Computers, he believes, will free us from the processing constraints of our biological brains, opening up the frontiers of human knowledge. They are, though, not yet as powerful as the human brain, which carries out around 100 to 10,000 trillion instructions per second. Nor may we ever be able to construct a silicon machine that could work in the same way as a human brain, for example, handling the frequent ambiguities of life.[88] However, Kurzweil believes this watershed could be reached by 2025. There might not be a single grain of sense in all these incredible speculations, but it is grains of silicon that have caused them. Using readily available sand, the human mind has created something that might one day exceed itself. But we need to be very careful not to exaggerate the potential of current technologies. As Gordon Moore said: 'I do not see an end in sight [to Moore's law], with the caveat that I can only see a decade or so ahead.'[89] So what can we expect in the next decade? At Intel, one silicon innovation is already coming to fruition.

Silicon photonics

In May 2012, Mario Paniccia, a researcher at Intel, led me through a maze of office cells at the company's headquarters in Santa Clara, California. We reached his desk, the same size as all the others, in the far corner of the room. 'The bosses get the windows,' he explained. While I waited for him to find the small silicon device I had come to see, I looked at the photographs on his wall. Next to the portraits of his family and friends, Paniccia was pictured several times with a man I recognised from the ten years I had spent on Intel's board: Gordon Moore. 'We finally got to play the full eighteen holes,' he said pointing to a picture of him standing side by side with Moore, dressed in golfing gear. Moore has clearly been an inspiration to Paniccia, one of a new wave of innovators pushing the limits of silicon technology. His current research is attempting to extend the exponential trend of Moore's law to the technology of data transmission. Having at last found what he was looking for, he opened a box and took out two small silicon chips connected with a cheese-wire-thin translucent optical fibre. 'This,' he said, 'is the future of communication.'

Optical fibres can carry more data faster and further than copper wires, yet they are rarely used. The fibre itself is made from glass, and so is cheap, but the lasers used to generate the light signals sent down the fibre are expensive. So, too, are the light boosters placed periodically along the fibre and the light decoders at the receiving end. These components cannot be mass-produced, and the system cannot be mass-assembled, making silicon communication a costly alternative to copper wires. The use of optical fibres has generally been restricted to the giant 'information highways' that link continents and countries, and which every second carry tens of terabytes of data (about one hundred times the information stored on your computer hard drive). Each connection can cost hundreds of millions of dollars.

Paniccia believes all that is about to change. The tiny 'silicon photonics' device he handed me could soon make high-performance optical fibres affordable everywhere, from big data centres to personal computers.[90] Reminiscent of Jack Kilby and Robert Noyce's breakthrough with the integrated circuit, the way to do this is to make the equipment almost entirely of silicon. Doing so enables the entire optical communication system to be mass-produced cheaply, building on fifty years of silicon manufacturing technology. Paniccia's device can carry 50 gigabytes every second, fast enough to download a high-definition movie in less than a second. His team is now working towards a one terabyte per second link, which could download the entire printed collection of the Library of Congress in about ninety seconds.

These silicon photonic devices are the latest inventions to be added to the complex infrastructure that supports our computation and communication needs. They combine silicon's interactions with light and electrons to produce a high-speed communication connection that can be cheaply mass-produced. Silicon has reinvented itself once again. This is the technology of tomorrow, but what about the day after? Just over the horizon one possibility seems particularly exciting, and it comes once again from carbon.

CARBON REVISITED: GRAPHENE

A new substance, looking like a piece of futuristic chicken wire, has the potential to be a miracle material for the twenty-first century, with the power to transform that might exceed silicon's. Unlike silicon, its development is not a story of glamorous entrepreneurs in sunny California, but one of pencils and sticky tape in the rainy north of England. At the start of the millennium, Russian-born professor Andre Geim and his student Konstantin Novoselov were working at the University of Manchester, investigating a new kind of transistor, made not from a semiconductor like silicon but from a conductor. They hoped to produce a device that was even smaller, faster and used less energy than anything currently available. They began to experiment with graphite, which consists of thin layers of carbon atoms stacked on each other, and from which pencil lead is made. As you write, pressure is applied to the tip of a pencil and the thin carbon layers of the graphite slide over each other to form the words on the page.

For many years, scientists had been investigating the unusual properties of structures made of pure carbon. Carbon's ability to bond with itself allows a great diversity of carbon molecules, including the long chains and rings that form the backbone of hydrocarbon fuels. In 1985, Harry Kroto's research team at Rice University in Houston, Texas, created a football-like cage of sixty carbon atoms called Buckminsterfullerene.[91] Several years later, hollow cylindrical carbon nanotubes became the 'wonder material' of the 1990s. Along the same lines, scientists wondered whether a thin sheet of carbon atoms could be made; most thought that a sheet would be unstable and crumple up when it got to the thickness of a single atom.

However, as they were investigating the properties of thin graphite layers, Geim and his student at Manchester made a startling discovery. Using ordinary sticky tape to peel flakes off a graphite block, they were able to create thinner and thinner sheets, reducing the thickness to only a few atoms. Eventually, as they looked through a microscope, they realised they had achieved what many had thought impossible: a sheet of carbon just one atom thick. They had made graphene. They began to explore the properties of this novel material, and the surprises continued. They discovered that it was the world's strongest material, three hundred times stronger than steel.[92] According to one calculation, it would take an elephant balancing on a pencil to break through layers of graphene the thickness of clingfilm. It combines that strength with very high flexibility and conductivity. It may be the world's best conductor of heat and electricity, better than both copper and silver, with virtually no electrical resistance at room temperature.[93] To top it off, it was also the most transparent material in existence. 'It was very unusual,' says Novoselov. 'Each time we worked with it we found something new and interesting: the optical properties, the electrical properties, the mechanical properties were all remarkable.'[94]

The initial findings were published in 2004 and they have continued to generate great scientific and commercial interest.[95] In 2010, only six years later, Geim and Novoselov were awarded the Nobel Prize in Physics. In its announcement, the Royal Swedish Academy of Sciences said: 'Carbon, the basis of all known life on earth, has surprised us once again.'[96] Carbon is the most versatile of the chemical elements. In the form of fossil fuels it developed the world, powering manufacturing, trade and communication; in the form of carbon dioxide it may transform the world again, permanently altering our climate and our way of life; in the form of graphene, it may revolutionise many of the products that make our lives better. Its transparency and conductivity could be used in solar cells and touch screens, its strength and flexibility in ships' hulls and space ships, and its semi-permeability in antibacterial bandages and water filters.[97] Lithium ion batteries made with graphene anodes might have ten times the storage capacity and recharge many times faster than current ones. Devices, such as phones, made with graphene transistors might be so thin that they could be rolled up and put behind your ear.

The potential for graphene is enormous but often the benefits of newly discovered materials are much exaggerated.[98] I am reminded of the technological optimism of the 1950s. Popular science magazines and comics predicted a future in which uranium would supply all our energy needs, heating homes, powering cars and even controlling the Earth's climate at the flick of a switch. Titanium, stronger, lighter and more corrosion-resistant than steel, was predicted to become as integral to modern life as iron. And graphene's sister material, Buckminsterfullerene, found few practical applications, and carbon nanotubes have yet to make a significant impact on industry.

Graphene has proved its potential in the laboratory. Whether it forms the basis of a revolution in products is a question of economics and engineering rather than science; commercialising a new material takes time, effort and money. It looks likely that its first commercial applications will be in thin-film touch screens and electronic paper, but for some of its more extraordinary applications we will probably have to wait for decades, if they arrive at all.[99] Graphene, though, is a perfect example of how the elements, probed with human curiosity and applied by human ingenuity, can surprise us time and time again, revealing new properties and powers that go on to transform the world.

Power, Progress and Destruction

I suspect that every generation in every place believes that their moment in history is moving more quickly than that of previous generations. They may be right. Today, everything, from technological advances to the growth in population, seems to be moving faster. Decisions are made rapidly, communicated more widely and their impact felt more broadly. What we do today has a far greater impact on humankind than what was done yesterday. Underlying what will be done is the way humankind uses the elements. Agricola cautioned us in the sixteenth century that 'good men employ them for good and to them they are useful. The wicked use them badly and to them they are harmful.'[1] I wonder what his practical and realistic advice to us today would be. Here is my view.

First, all of us need to be aware of not only the good things we get from using the elements, but also their negative consequences. You can see that in carbon and its impact on the global climate, or uranium and its use in a weapon of mass destruction. To foster an understanding of these dangers will take a great commitment to education. It will also take an enormous amount of communication, not least to overcome the voices of those with vested interests who want to be blind to the negative.

Second, while all those who have predicted that we will run out of one or another element, mineral or commodity have so far been wrong, they may eventually be right. We need to keep investing in the technologies that make our use of these exhaustible reserves more productive. And we should not prejudge which of these technologies to use but to decide on the evidence of their overall merits. Short-run considerations of supply and demand will not provide the right basis to do the research and development needed for the future. Leaders are needed to take the risk of looking to and acting for the long run.

Third, we need to take a fresh look at the age-old human characteristic of greed. The seven elements have always inspired greed; their utility and power appeal to the self-interest in all of us. Many people are captivated by the idea of being rich and are prepared to do terrible things to get there. They have fought to control land, they have knowingly polluted, they have exploited labour and they have used their wealth to do the same all over again.

Greed cannot be eliminated, but it can be controlled and directed for good. Society can make laws that forbid exploitation of people and the environment. Around the world, where the law is enforced, companies or individuals buy land rather than steal it, employ workers rather than enslave them, and protect ecosystems rather than destroy them. The law does this by prohibiting the use of force, compelling people to act within defined boundaries. Among other things, the good laws create a marketplace, founded on mutual consent, and thereby align self-interest with the interests of all involved. They place a harness on greed, directing its force to the service of humanity.

In most cases, the law takes effect through regulations, which, at their most basic level, prohibit activities that are harmful to society. Society forbids the use of iron swords for murder or the use of slaves in the pursuit of gold. But often what is needed requires a more complex construction. Our use of carbon energy to raise the living standards of billions has come about hand in hand with the loss of human life and pollution of our land, air and water. Effective regulation requires a balancing act between providing the energy needed for economic development and minimising the harm to people and the environment.

Today, it is most difficult, and indeed most important, to achieve this balance in making regulations to reduce the risk of climate change. Governments have to think carefully about how they design the rules of the game; they cannot simply forbid the use of carbon fuels. Developed economies would crash and developing economies would stand still. Rather, they must design mechanisms that direct self-interest towards reducing energy use, decarbonising energy production and capturing carbon. The theoretical ideal is a global carbon tax that forces polluters to take account of the damage caused by carbon dioxide. In practice, international and domestic politics make that impossible and so we must make do with a complex collection of subsidies, regulations, taxes and carbon pricing. It is a messy and inefficient process, but it is society's way of curbing carbon's potentially destructive potential. I am optimistic that it will work eventually.

Intelligent regulation not only protects citizens and the environment but also companies. For example, bad practice, by a few operators, in the way in which wells were completed to produce gas from shale led to problems that gave the industry a bad name. Regulation must prevent the few from damaging the many as well as encourage competition. Only when the powerful grip of Standard Oil was removed with its dissolution in 1911 could other US oil companies begin to thrive. Regulation was used to restrain the greed of Standard Oil while fostering the aspirations of smaller market players. I saw a similar situation in Russia in the 1990s, where bribes, threats and fraud were the norm in business. The legal system had been twisted by a powerful few for their own benefit. There were many rules, but they were applied selectively in the interests of those with political power.

There remain many other countries in which society is left stranded as the apparatus of the state turns a blind eye to, or, worse still, is complicit in, the destructive extraction of the elements. Africa is the home of vast and untapped resource wealth, yet much of that revenue is vulnerable to corruption, and ends up in the pockets of the elite. As I look back, it is clear to me that the most powerful force for change in those countries is to push governments, by applying international pressure, to disclose payments from resource development and their general expenditures. That would allow citizens to hold them to account. And it would compel change in the enforcement of internal laws designed to prevent corruption.

Through my involvement in the Extractive Industry Transparency Initiative I have seen how the bright light of transparency can ensure that oil wealth reaches the hands of the citizens to whom it belongs. Across the world of business, transparency is becoming less of an option. More instant and extensive communication enables citizens and NGOs to observe almost every activity of a business, to rally opposition against it and to launch powerful global campaigns very quickly at virtually no cost.

Laws today are stronger, and the greed of individuals better confined, than at any point in history. As societies develop, aided by the force of transparency, they demand better working practices and a greater respect for the environment. Transparency does not solve everything, but where it is practically applied it seems to be making a difference.

Fourth, we need to celebrate philanthropy. Laws are a vital mechanism to direct the use of the elements for the good of humanity. This, however, is not enough. Even when the elements are used to foster human progress and prosperity, they can also create great inequality. We saw this in the stories of Carnegie and Rockefeller. But by giving away the fortunes they had made in the production of steel and oil, they went some way in reducing the gap between rich and poor in the society of their age. The evidence suggests their philanthropy was driven by a series of motives. Clearly, they wanted to leave a legacy, to be immortalised with their names associated with great institutions like Carnegie Hall and Rockefeller University. They recognised that great businesses are transient in comparison with great institutions. They were also probably motivated by guilt. Carnegie and Rockefeller were both criticised during their lifetimes as 'robber barons' who trampled on competition and mistreated their workforce. I doubt they would have been so generous were it not for the scandals of the Homestead strike and the Ida Tarbell exposé.

Finally, they were motivated by a paternalistic idea of compassion. They believed in their own ability to build a better society, and to improve the lot of the working class, by applying not just their capital but their own ideas. The Rockefeller Foundation and the Carnegie Corporation of New York both still hold more than US $2.5 billion; together they make annual donations of hundreds of millions of dollars across the globe. For good reason, much of that great philanthropy has been to education, the purest form of social investment, improving the stock of human capital to aid society and allowing each individual to fulfil their own potential.

In the twentieth and twenty-first centuries the profile of the nature of philanthropy has changed. Some believe, especially in Europe, that the state, with its power to tax and spend, is the best mechanism for correcting inequality and promoting progress. Others, including some of the world's richest people, are thinking differently and leading by example. Bill Gates and Warren Buffett have pledged to give away more than half their wealth, and are persuading billionaires across the US to do the same. Gates's words evoke Carnegie's sentiments expressed almost a century ago: the rich must give away more money to causes that benefit the many and that are managed well.

Gates and Buffett are in the vanguard of creating positive change through contemporary philanthropy. Change in society has always depended on the examples set by great leaders, individuals who have a clear vision. Those individuals take personal risks since their actions are likely to be judged both good and bad by different people at different times. General Groves, on taking the helm of the Manhattan Project in 1942, was told by his superior in the US army: 'If you do the job right, it will win the war.'[2] Three years later, the atomic bomb created under his leadership destroyed a city and, as predicted, ended the war in the Far East.

Leaders must be able to change their own vision as society changes around them. When Groves and Oppenheimer saw the terrifying impact of the weapon they had created, their purpose and direction changed. They saw the need to direct uranium's extraordinary energy for the benefit, rather than the destruction, of humankind. The role of leaders in stemming the spread of nuclear weapons remains vitally important today. The sheer potency of nuclear weapons has made political control of them almost impossible. Putting the lid back on Pandora's Box has proved such an intractable problem because it requires cooperation among nations, when each has the incentive to betray. We would all be better off in a world without nuclear weapons, but for each individual nation it is better to have this potent weapon than not. There is no easy solution to this sort of dilemma: as long as each of us acts selfishly and rationally, we reach a result that is bad for us all.

The same dilemma faces us in our use of carbon-based energy: we would all be better off if carbon emissions were reduced, but for each individual and each nation it is better to consume more energy than less. In these complex cases, it seems simultaneously frustrating to see inaction, but futile to seek solutions. What is required is good leadership: to cooperate globally, to move beyond self-interested rationality and to take unilateral action towards the common good. Only great leaders can forge a path to a better future, whether free from nuclear weapons or from the risk of climate change. We need brave and inspirational individuals who are prepared, in a sense, to act irrationally: to make a sacrifice with no guarantee of reciprocity.

Leaders such as George Shultz, the former US Secretary of State, and Hidehiko Yuzaki, the Governor of Hiroshima, are working towards a nuclear weapon-free future. Their work is focused on educating people about the terrifying consequences of a nuclear explosion as well as on negotiating treaties. Only when I visited Hiroshima did I comprehend the full extent of human hurt in the atomic explosion and the imperative to stop a similar event ever happening again.

In the case of climate change, some countries are already taking the lead towards a carbon-free future. Government support in the US, China and Germany has catalysed the growth of domestically generated clean energy at an astonishing rate. Economies of scale have been achieved and as a result costs have plummeted. Global solar capacity, for example, has grown annually at an average rate of 60 per cent over the last five years, while the cost of solar cells has fallen by three-quarters. Perhaps one of the largest near-term reductions in the world's output of carbon dioxide will come from the greater proportionate use of gas, notably from shale. George Mitchell, who relentlessly pursued the idea of shale gas, was the eternally optimistic entrepreneur who made this possible.

Inertia remains among many people who see climate change as an uncertain and far-off risk. It will take more great leaders to embrace new technology, to imagine a society no longer so dependent on carbon and to put that case to the people.

Great leadership, though, is not just about solving global challenges. It is also about understanding society's day-to-day needs and taking the risk to back the new and novel. George Eastman did not invent photography and Henry Ford did not invent the automobile, but they both had the vision to see that these world-changing technologies could be turned into ubiquitous, affordable products.

Steve Jobs' vision was to change the world through easy-to-use computing technology. When the computer was the tool of big business, he made it the intuitive tool for the people. He persistently pursued his idea of perfection, with a passion that was contagious. True to his character as a leader, Jobs created sophisticated but simple devices that changed the lives of billions. He followed in a long line of entrepreneurs who took great risks to establish the computer industry. William Shockley first saw the potential of the silicon transistor and moved to the West Coast to found his own semiconductor company. When his staff stopped believing in the direction he was taking the company, they left and founded their own. Out of this came Intel, which, under the leadership of Gordon Moore and Andy Grove, became the world's largest manufacturer of advanced computer chips and a household name today. The work of these people has allowed a new form of change to come to the fore. The Arab Spring resulted from a reaction to the loss of freedom and dignity, but it was enabled by silicon-based mobile devices. I doubt whether the silicon chip pioneers, Jack Kilby and Robert Noyce, would ever have believed this to be possible.

The great progress and prosperity that humanity has achieved through our use of the elements is driven by people – scientists, business people and politicians – who are leaders in the broadest sense. They have all looked to the future and aspired to create something better. The modern pace of innovation is so fast that few can imagine how the elements will change the world in the next century. Who could have envisaged the possibilities for uranium or silicon only seventy years ago? It is those leaders with a vision, who take risks and who challenge the status quo, who will unleash the full potential of the elements and direct the next chapter of their use for the benefit of humanity.

ACKNOWLEDGEMENTS

While I was finishing this book in Venice, I was given a copy of Walter Benjamin's essay on collecting books.[1] He tells the story of a poor school teacher who acquired a large collection of works by writing them all himself, having seen the titles in book catalogues but been unable to afford them. This he does to make the point that the noblest way of collecting books is to write them oneself.

Benjamin's essay made me realise that I am a book collector in more than one way. He lists, in order of worth, three other ways of collecting books: borrowing from a collector with taste, buying from a dealer and buying at auction. That gives me a chance to thank those who have helped with the three antique Italian books referred to in *Seven Elements*: my auction representative and book dealer, Robin Halwas, who not only helped me acquire my wonderful copy of the first edition of Biringuccio (1540), whose known provenance commences with the French royal physician François Rasse des Neux in 1552, but also the first edition of Brustolon's engraved views of Venice derived from drawings and paintings by Canaletto; and Kristian Jensen of the British Library who gave me access to the copy of Agricola owned and annotated by Prince Henry, the son of King James I, after I had failed miserably to buy an inferior copy at auction. I am grateful to both of them and to the many others who continue to inspire my collecting.

This book has been enormously helped by the advice of several friends who have given freely of their time. I owe my thanks to Daniel Yergin, the Pulitzer prize-winning author of *The Prize* and *The Quest*; Dr David Allen, formerly a director of BP; Professor Lord (Peter) Hennessy, a distinguished historian of modern and contemporary Britain; Ian Davis, formerly Senior Partner of McKinsey; Donna Leon, the novelist; Laurence Hemming, philospher and author; Ernst Sack and Simon Maine, members of the Riverstone team; Philippa Anderson, my collaborator in writing my last book, *Beyond Business*; Lady Romilly McAlpine, my oft-time lunch partner in Venice; Nick Butler, formerly a senior colleague at BP; and David Rawcliffe and Matthew Powell, my former and present personal research assistants.

My thanks go to Charles Merullo, managing director at Endeavour London, for his assistance in obtaining many of the photographs reproduced in the book.

The germ of the idea for this book came from my publisher, Alan Samson, who also gave me great guidance on earlier drafts. Ed Victor, my literary agent, was, as always, a great help in many ways. I thank them both.

Thomas Lewton spent a year of his life thoroughly researching all the material for this book. He did this with patience and style, never blinking when asked to hit a deadline. I thank him deeply for his invaluable work.

As always, my thanks go to my partner, Nghi Nguyen, who provided invaluable guidance in the final editing.

Finally, I would like to thank all those who gave me the experiences on which to write this book: my former colleagues at BP and its present chief executive, Bob Dudley; and David Leuschen and Pierre Lapeyre, the founders of Riverstone who have provided me with continued challenge and interest in the energy industry and the world around it.

John Browne,
Venice, September 2012

LIST OF MAPS

All maps drawn by John Gilkes

LIST OF ILLUSTRATIONS

Frontispiece © Ars Thanea *www.arsthanea.com.*

THE ESSENCE OF EVERYTHING

1. 29 December 1931: English physicist Sir William Henry Bragg surrounded by school children after giving a lecture on the history of light at the Royal Institution. *Getty Images.*
2. Prince Henry's copy of Agricola's 1556 masterpiece *De re metallica.* © *British Library Board.*

IRON

3. Currier and Ives' depiction of the Battle of Hampton Roads, 1862. *Author's collection.*
4. Photograph taken in September 1937 of Adolf Hitler, Italian Fascist state's head Benito Mussolini and Krupp steel corporation's director Gustav Krupp, during their visit to the Krupp factory in Essen. *AFP/Getty Images.*
5. *Thunder Horse* aboard the MV *Blue Marlin. Courtesy of BP plc.*
6. An engraving showing workers in the Carnegie Steel Works as they convert iron into steel using the Bessemer process. Pittsburgh, Pennsylvania, 1886. *Getty Images.*
7. Pig iron is poured into a Bessemer Converter during the final process of steel production at ThyssenKrupp AG's plant in Duisburg, Germany, 2012. *Bloomberg/ Getty Images.*
8. Bessemer and Carnegie meet at the Institute of Materials, Minerals and Mining in London. *Author's collection.*
9. Gerald Kelly's 1924 posthumous portrait of Henry Clay Frick, which hangs in the West Gallery of the Frick Collection, New York. *Courtesy of The Frick Collection/ Frick Art Reference Library Archives.*
10. Carnegie Hall in Midtown Manhattan, New York, in the year of its opening, 1891. *Getty Images.*

11. The Iron Lion of Cangzhou, China. *Beijing Energy Club.*
12. Stieglitz's iconic photograph of the Flatiron Building, 1902–1903 photogravure (32.7 cm x 16.83 cm). *San Francisco Museum of Modern Art, Alfred Stieglitz Collection, Gift of Georgia O'Keeffe.* © *Estate of Alfred Stieglitz.*
13. A postcard from circa 1907 shows the Flatiron in popular culture. *John Cline, Timefreezephotos.com. Original postcard by the E. L. Theocrom company.*

CARBON

14. A view of the fireworks during Venice's Festa del Redentore. *Author's collection.*
15. Naft Safid Rig 20, the site of one of the world's greatest oil well blowouts in 1951. *Author's collection.*
16. Edouard Boubat's photograph of London Bridge, 1958. *Gamma-Rapho/Getty Images.*
17. Kunming Lake at the Summer Palace, Beijing, China, in 2008. *Tim Graham/Getty Images.*
18. The author with Henry and Bill Ford. *Author's collection.*
19. Agricola's bituminous spring – Prince Henry's copy of Agricola's *De re metallica.* © *British Library Board.*
20. Oily Rocks, near Baku, Azerbaijan, depicted on a Soviet postage stamp in 1971. The paths and platforms can be seen winding into the distance.
21. Oily Rocks, Baku, Azerbaijan. *Courtesy of SOCAR and BP plc.*
22. The *Exxon Valdez* disaster: workers train their hoses on oiled rocks at McPherson Bay on Naked Island as a perforated hose jets water in the foreground. April 1989. *MCT/Getty Images.*
23. A fire burns on the North Sea oil rig *Piper Alpha* after an explosion in July 1988 which killed more than 160 people. *Press Association.*
24. Portrait of John D. Rockefeller Sr, painted by artist John Singer Sargent in 1917. *Time & Life Pictures/Getty Images.*
25. Doing business in Russia. The author (third left) says farewell to Putin (third right). Also shown is Mikhail Friedman (front right). *Author's collection.*
26. Cars queuing at a petrol station on Woodford Avenue, London, during a petrol shortage, December 1973. *Getty Images.*
27. Burning oil wells in Kuwait, set afire during Gulf War, March 1991. *Time & Life Pictures/Getty Images.*
28. The birth of the supermajors: at the New York Stock Exchange on the day of BP Amoco's listing. From left to right: Dick Grasso (CEO of the NYSE), the author, Maria Bartiromo (CNBC journalist), Larry Fuller (co-chairman of BP Amoco) and Rodney Chase (deputy CEO of BP Amoco). *Author's collection.*

29. A BP meeting in Venezuela, 1980. Note the author and his hair at front right, along with the cigars and wood panelling. *Author's collection.*

30. The author with the Prime Minister of Trinidad and Tobago at the time, Patrick Manning, January 2007. *W. Garth Murrell/PIPS Photography.*

31. LNG trains in Trinidad and Tobago, operated by Atlantic LNG and part-owned by BP. *Courtesy of BP plc.*

32. A fracking spread in Lancashire, where Britain's most promising shale gas reserves are to be found. *Cuadrilla Resources Ltd.*

33. The author delivers his speech on addressing climate change at Stanford University, May 1997, watched by the then Dean of the Stanford Graduate School of Business and Nobel prize-winning economist Mike Spence. *BP and Stanford University Graduate School of Business.*

34. Solar panels with Al Gore (BP Solar, California, 1998). *Author's collection.*

35. Discussing climate change in California with British Prime Minister Tony Blair and Governor Arnold Schwarzenegger, July 2006. *Author's collection.*

GOLD

36. The gold raft of El Dorado, on display in the Museo del Oro, Bogotá, Colombia. *Banco de la República, Bogotá – Gold Museum Collection.*

37. A *poporo* – my own, purchased in the early 1990s. *Author's collection.*

38. Brustolon's engraving, c. 1768, of the inauguration of the Doge, based on a drawing by Canaletto. *Author's collection.*

39. A Venetian ducat from the reign of Doge Andrea Gritti, 1523–1538. © *The Trustees of the British Museum. All rights reserved.*

40. A gold coin, suspected to have been minted during the reign of the fabulously rich King Croesus, ruler of the Lydians from approximately 560–547 BC. © *The Trustees of the British Museum. All rights reserved.*

41. The Incan emperor Atahualpa (1497–1533) executed by the Spanish conquistadors by garrotte. Artist unknown, 1754. *Getty Images.*

42. Sutter's Mill Nugget found by James Marshall began the California Gold Rush. The nugget is kept at the Smithsonian Institute, Washington, DC. *Time & Life Pictures/Getty Images.*

43. Chinese workers panning for gold in California. A man in a coolie hat digs as another man kneels and sifts, c. 1855. *Getty Images.*

44. Newton's invention to stop clipping survives in the inscriptions on British £1 and £2 coins. *Author's collection.*

45. Kennecott Utah Copper's Bingham Canyon Mine, owned and operated by Rio Tinto. *Rio Tinto.*

46. A view of the Serra Pelada mine in Brazil. © *Sebastião Salgado/NB Pictures.*

SILVER

47. Silver smelting depicted in Prince Henry's copy of Agricola's *De re metallica*, 1556. © *British Library Board.*

48. Virgin Cerro, anonymous, eighteenth century. Unframed painting is 140 cm x 107 cm, and is hung in the national mint of Potosí, Bolivia. *National Mint BCB Cultural Foundation, Potosí, Bolivia.*

49. Fox Talbot's calotype: two men positioning a ladder up against a loft. Original Publication: *The Pencil of Nature – The Ladder*, 1844. *Getty Images.*

50. Daguerreotype made in 1838 by Louis-Jacques-Mandé Daguerre of a Paris boulevard, the first photograph to show a human being. *Topfoto.*

51. Illustration on Kodak Brownie camera box, c. 1910. *SSPL/Getty Images.*

52. View from the author's home in Venice. *Author's collection.*

53. The Singaporean Botanical Gardens, with my mother, 1953. *Author's collection.*

54. South Vietnamese National Police Chief Brig. Gen. Nguyen Ngoc Loan executes a Viet Cong officer with a single pistol shot in the head in Saigon, 1 February 1968. Carrying a pistol and wearing civilian clothes, the Viet Cong guerrilla was captured near Quang Pafgoda, identified as an officer and taken to the police chief. *AP Photo/Eddie Adams/Press Association*

55. Jewish children, survivors of Auschwitz, behind a barbed-wire fence, Poland, February 1945. Photograph taken by a Russian photographer during the making of a film about the liberation of the camp. The children were dressed up by the Russians with clothing from adult prisoners. *Getty Images.*

56. The author's Persian silver boxes. *Author's collection.*

57. From left to right: Vorochilov, Molotov, Stalin pose at the shore of the Moscow-Volga Canal, in 1937 in this manipulated picture. In the original picture Nikolai Yezhov was standing on the right. Yezhov was the senior figure in the NKVD (the Soviet secret police) under Joseph Stalin during the period of the Great Purge. After Yezhov was tried and executed his likeness was removed from this image. *AFP/Getty Images.*

58. The Hunt brothers are sworn in before a Congressional subcommittee investigating the collapse of the silver market, February 1980. *Corbis.*

URANIUM

59. Hiroshima's A-Bomb dome, taken by the author on a visit in 2012. *Author's collection.*

60. 'Weakly Writhing' by Tomomi Yamashina, who was sixteen at the time of the bomb and standing 3,600 metres from the hypocentre in front of the Hiroshima

First Army Hospital. Hiroshima Peace Memorial Museum A-bomb Drawings by Survivors. *Author's collection.*

61. Nuclear physicist Robert Oppenheimer (left) with Major General Leslie Groves, by the remains of the tower from which an atom test bomb was ignited, at Los Alamos, California, 1943. *Getty Images.*

62. Her Majesty Queen Elizabeth II opens Calder Hall, 1956. *Nuclear Decommissioning Authority.*

63. Captain Atom – *DC Comics.*

64. The operational Blue Steel stand-off bomb carried Britain's nuclear deterrent between 1964 and 1975. The Science Museum's Blue Steel shown here is a test vehicle and is fitted with a Double Spectre rocket engine. The engine is being examined by a Science Museum curator prior to being photographed for a Science Museum book on the Black Arrow Rocket. *SSPL/Getty Images.*

65. This satellite view shows the Fukushima Dai-ichi Nuclear Power plant after a massive earthquake and subsequent tsunami on 14 March 2011 in Futaba, Japan. *DigitalGlobe/Getty Images.*

66. Pakistan's top scientist Abdul Qadeer Khan addresses a gathering after inaugurating the model of the country's surface-to-surface Ghauri-II missile, in Islamabad, 28 May 1999. *Usman Khan/AFP/Getty Images.*

67. The author with Governor Yuzaki of Hiroshima, 2012. *Author's collection.*

TITANIUM

68. SR-71B Blackbird aerial reconnaissance aircraft photographed over snow-capped mountains in 1995. *Getty Images.*

69. A titanium Project 705 Soviet submarine, designated Alfa class by NATO. *Ministry of Defence.*

70. Isaac Newton using a prism to break white light into spectrum with Cambridge roommate John Wickins. Engraving from 1874. *Getty Images.*

71. The Guggenheim Museum, Bilbao, Spain. *Allan Baxter/Getty Images.*

72. A view of Lake Tio overlooking the Tio ilmenite mine operated by Rio Tinto Fer et Titane (formerly QIT Fer et Titane). Quebec, Canada. *Rio Tinto Fer et Titane.*

SILICON

73. An image from a section on glassmaking in the author's copy of Biringuccio's *De la Pirotechnia. Author's collection.*

74. The author's glass elephants. *Author's collection.*

75. Glass blowing in Murano, 2010. *Author's collection.*

76. The Hall of Mirrors in Versailles, 2012. *Author's collection.*

77. Glass petrol pump globes by Chance Brothers. *With thanks to Michael Joseph.*

78. Glaziers, painters and decorators posing in front of the Great Conservatory at Chatsworth House, early 1900s, following the completion of extensive storm damage repairs. Photographer and date unknown. © *Devonshire Collection, Chatsworth. Reproduced by permission of Chatsworth Settlement Trustees.*

79. A group photograph of some of those responsible for building the Crystal Palace for the Great Exhibition of 1851 in Hyde Park, London. The two men on the beam (top) are iron-fitters. *Getty Images.*

80. William C. Miller, resident photographer and pioneer in astronomical photography, at the Mt Palomar Observatory. *Time & Life Pictures/Getty Images.*

81. The picture depicts the 6.8 MW solar photovoltaic plant owned by Soemina Energeia S.r.l. (owned directly by AES Sole Italia S.r.l., the Italian subsidiary of AES Solar). It is located in the Municipality of Ciminna, province of Palermo, Italy. *Reprinted with permission of Antonio Nastri and AES Solar.*

82. Pages 11 and 12 from Da Vinci's notebook. The notebook was not originally a bound volume, but was put together after Leonardo's death from loose papers of various types and sizes. It can be found in the British Library catalogued as Arundel 263n ff.86 v-87. © *British Library Board.*

83. An operator with an IBM 1130. *Courtesy of International Business Machines Corporation.*

84. Shockley, Bardeen and Brattain's point-contact transistor from 1947. Public announcement of the transistor was made on 1 July 1948 by Bell Laboratories. In 1956, Shockley, Bardeen and Brattain shared the Nobel Prize for their discovery of the transistor effect. *Reprinted with permission from Alcatel-Lucent USA Inc.*

85. A 22nm transistor, 2011. © *Intel Corporation.*

86. Gordon Moore and Andy Grove, 1990. © *Intel Corporation.*

87. The author with Mario Paniccia, the man behind silicon photonics, in May 2012. *Author's collection.*

88. An electron microscope image of individual carbon atoms in a sheet of graphene, the thinnest and strongest material known to science. *MCT/Getty Images.*

NOTES

Preface

1. Every atom consists of a nucleus, made of protons and neutrons, which is orbited by electrons. Atoms of the same element have the same number of protons in the nucleus. As you move from left to right along the 'periods' of the periodic table, the number of protons in the nucleus increases by one with each step. So, too, does the number of electrons, which is the same as the number of protons, and which orbit the nucleus in larger and larger 'shells' as their number increases. Elements in the same columns, or 'groups', of the periodic table have similar chemical properties because of similarities in the arrangement of electrons, neutrons and protons.
2. Fossil fuels are produced from dead plants and animals when subjected to intense heat and pressure over a long period of time.

The Essence of Everything

1. W. H. Bragg, *Concerning the Nature of Things* (London: G. Bell and Sons Ltd, 1925).
2. By directing x-rays at regular crystal structures, the Braggs were able to determine their atomic structure from the angles and intensities at which these x-rays were reflected. The Braggs won the 1915 Nobel Prize in Physics for their work.
3. Bragg, *Concerning the Nature of Things*, p. 6. John Dalton, the English chemist and physicist, hypothesised the existence of atoms of different elements distinguished by their weight. In his theory, presented to the Royal Institution in 1803, these atoms could not be divided, created or destroyed but could be combined or rearranged in simple ratios to form compounds. Later in the nineteenth century, Dmitri Mendeleev, a Russian chemist, noticed a pattern in the chemical properties of the known elements. The weights of elements with similar properties increased regularly. This periodicity led Mendeleev, in 1869, to arrange the elements in rows or columns according to atomic weight; he would start a new row or column when

the chemical behaviour began to repeat. By leaving gaps in the table when no known element would fit, Mendeleev was able to use his periodic table to predict the properties of undiscovered elements.

4. Ibid., pp. 186–7.

5. John Browne, *Beyond Business* (London: Weidenfeld & Nicolson, 2010), Chapter 10.

6. Ibid., Chapter 9.

7 Ibid., Chapter 6.

8. The 'Great Leap Forward' was coined by Jared Diamond in *Guns, Germs and Steel* (London: Jonathan Cape, 1997), p. 39.

9. In *The Meaning of It All*, Feynman writes that science has value because it provides the power to do something. He asks: 'Shall we throw away the key and never have a way to enter the gates of heaven? Or shall we struggle with the problem of which is the best way to use the key? That is, of course, a very serious question, but I think that we cannot deny the value of the key to the gates of heaven.' (London: Penguin Books, 1998), pp. 6–7.

IRON

1. Report of flag officer Franklin Buchanan, C. S. Navy. Naval Hospital, Norfolk, VA, 27 March 1862, in Mills, Charles, *Echoes of the Civil War: Key Documents of the Great Conflict* (BookSurge Publishing, 2002), p. 118.

2. Thomas Oliver Selfridge Jr, in David Mindell, *Iron Coffin: War, Technology and Experience Aboard the USS Monitor* (Baltimore: Johns Hopkins University Press, 2002), p. 71.

3. Navy Secretary Welles in Mindell, *Iron Coffin: War, Technology and Experience Aboard the USS Monitor*, p. 72.

4. Currier & Ives was a nineteenth-century New York-based print-making firm. It was one of the most successful and prolific lithographers in the US. The lithograph is inscribed, somewhat inaccurately: 'The Battle of Hampton roads ... In which the little Monitor whipped the [*Virginia*] and the whole school of rebel steamers.'

5. In the lithograph, the USS *Virginia* is referred to as the 'Merrimac', after the sunken ship from which the *Virginia* obtained its frame.

6. William Harwar Parker, *Recollections of a Naval Officer, 1841–1865* (Annapolis: Naval Institute Press, 1985), p. 288.

7. G. J. Van Brunt to Welles, Mar. 10 1862, in Mindell, *Iron Coffin: War, Technology and Experience Aboard the USS Monitor*, p. 74.

8. Only the *Monitor*'s captain was injured during the battle when he was blinded by a shell exploding into his eyes through a slit in the armour.

9. Mindell, *Iron Coffin: War, Technology and Experience Aboard the USS Monitor*, p. 1.

10. Mary Elvira, Weeks, *Discovery of the Elements* (Kessinger Publishing, 2003; first published as a series of separate articles in the *Journal of Chemical Education*, 1933), p. 4.

11. Otto von Bismark, *Blut und Eisen*, 1862.

12. This was the Schwerer Gustav cannon, named after Gustav Krupp, weighing 1,350 tonnes and with a 32-metre-long barrel. In the siege of Sevastopol in June 1942, 'Gustav' wreaked destruction on the Soviet naval base. One shot penetrated through 30 metres of earth, detonating inside an underground ammunition store.

13. The return of Alsace-Lorraine doubled the theoretical steel capacity of France, but it struggled to reach this potential. Shortly after the end of the First World War, France suffered a coal shortage as a result of the destruction of mines and wearing out of rails during the war.

14. Hitler said: 'The forty-eight hours after the march into the Rhineland were the most nerve-racking in my life. If the French had then marched into the Rhineland we would have had to withdraw with our tails between our legs, for the military resources at our disposal would have been wholly inadequate for even a moderate resistance.' R. Manvell and H. Fraenkel, *Adolf Hitler, The Man and the Myth* (New York: Pinacle, 1973), p. 141.

15. On the arrival of British and America troops in April 1945 they found that the city had been completely destroyed. To prevent the rise of the Ruhr armament factories, the Allied forces broke up what machinery was left and sent it to neighbouring countries as war reparations. 'No Krupp chimney will ever smoke again,' wrote the accountant put in charge of the Krupp factories. As for the Krupps, Alfred Krupp was imprisoned for war crimes and stripped of his wealth, just as his father, Gustav Krupp, had been after the First World War, but was soon released and his property returned to him. Peter Batty, *The House of Krupp* (London: Secker & Warburg, 1966), p. 12.

16. Robert Schuman, *The Schuman Declaration*, 9 May 1950.

17. Ibid.

18. In 1957, the Treaty of Rome created the European Economic Community (EEC), or 'Common Market'. The Maastricht Treaty, signed in 1992, set the EU on the path towards the euro, launched in 1999 and used today by seventeen of the EU's twenty-seven members.

19. In his memoirs, Jean Monnet, a chief architect of the ECSC, writes: 'Coal and steel were at once the key to economic power and the raw materials for forging the weapons of war. This double role gave them immense symbolic significance, now largely forgotten ... To pool them across frontiers would reduce their malign prestige and turn them instead into a guarantee of peace.' Jean Monnet, *Memoirs* (London: Collins, 1978), p. 293.

20. In October 2012 I visited the Zeche Zollverein Coal Mine Industrial Complex in

Essen for a meeting of the Accenture Global Energy Board, which I chair. Zeche Zollverein is now a museum and a UNESCO World Heritage Site, and its iconic shaft 12, built in distinctive Bauhaus style, is known as 'the most beautiful coal mine in the world'. Zeche Zollverein is typical of the transformation of Western economies from primary and secondary to tertiary industries: having been taken on a guided tour of the coal mine museum, I then spent the afternoon in a conference room on the top floor of the museum.

21. Unity is, however, far from apparent among member states of the European Union today. The global financial crisis of the later 2000s has led to a crisis of confidence among European member states, threatening the global recovery. Political and economic unions are always susceptible to break-up as, in times of stress, national interests strengthen.

22. Half of all iron consumed is used for construction purposes, a quarter is used to make machinery and a tenth to make automobiles. Only 3.3 per cent of iron is used in the oil and gas industry.

23. A detailed account of this period of BP's history can be found in J. H. Bamberg, *The History of the British Petroleum Company*, Vol. 2: *1928–1954* (Cambridge: Cambridge University Press, 1994), pp. 206–29.

24. The previous largest semi-submersible platform, used to tap the Åsgard oilfield in Norway, had a displacement of 85,000 tonnes (compared to 130,000 tonnes for *Thunder Horse*).

25. 'Building The Big One', *Frontiers*, April 2005, www.bp.com

26. In *Concerning the Nature of Things*, Bragg writes that, in steel, carbon atoms 'are forced into the empty spaces between [iron atoms]. We can easily see that this may distort the iron crystal, and prevent the movement along a plane of slip', p. 226.

27. Bessemer's resolve to produce a superior metal was furthered by his invention of rotary action bullets, the circular motion of which improved their range and accuracy. He had made a small cast-iron mortar to prove his concept by firing shells within the grounds of his Baxter House home, but the cast-iron cannons were too weak to withstand the pressure resulting from the heavier projectiles and often broke.

28. Memoir written in 1890. Joseph Needham, *Science and Civilisation in China*, Vol. 5, Part 11 (Cambridge: Cambridge University Press, 2008), pp. 361–2.

29. Henry Bessemer, *Sir Henry Bessemer, F.R.S.: An Autobiography* (London: Office of Engineering, 1905), pp. 143–4.

30. Precursors to Bessemer's process exist, such as the 'air boiling' process used by William Kelly in the US to make iron railings years before Bessemer's invention. However, Kelly did not create a liquid product using only air; his was an extension of an existing process, rather than an entirely new process. As far back as the eleventh-century Song Dynasty in China there are records of partial decarbonisation using a cold-air-blast technique.

31. Throughout the book, inflation is measured using the Consumer Price Index (CPI). www.measuringworth.com

32. Bessemer exhibited the first steel nails ever made at the International Exhibition of 1862. In America, where many houses were built from wood, steel nails dramatically reduced labour time as no holes needed to be bored before the nail was driven into wood. In his autobiography, Bessemer explains how young girls in the Black Country near Wolverhampton no longer had to work in smoky, grimy smithies shaping nails. He writes: 'I have often felt that if in my whole life I had done no other useful thing but the introduction of unforged steel nails, this one invention would have been a legitimate source of self-congratulation and thankfulness, in so far as it has successfully wiped out so much of this degrading species of slavery from the list of female employing industries in this country.' Bessemer, *An Autobiography*, pp. 378–9.

33. Krupp was informed of Bessemer's process by Richard Longsdon, a brother of Bessemer's friend and collaborator Frederick Longsdon. The new convertor was kept a secret in Krupp's works at Bessemer's request as he was unable to secure a patent for his design in Prussia. So as to disguise the new invention, Krupp's Bessemer works were named 'Wheelshop C'.

34. Bessemer had 117 patents to his name, 40 per cent of which were totally unrelated to the iron and steel industry.

35. Bessemer, *An Autobiography*, pp. 53–4.

36. Bessemer succeeded the Duke of Devonshire as its second President in 1871.

37. Such a fate is not uncommon among inventors of iron production processes. Dud Dudley was among the first Englishmen to smelt iron ore with coke, rather than expensive and increasingly scarce charcoal. He laid the foundations of many fortunes, but suffered a life of hardship. Henry Cort invented the puddling process for making iron and steel, but also ended his days a ruined man.

38. Carnegie was President of the Institute from 1903 to 1905.

39. Carnegie wrote: 'I am neither mechanic nor engineer nor am I scientific. The fact is I don't amount to anything in any industrial development. I seem to have a knack of utilising those that do know better than myself.' Bodsworth, *Sir Henry Bessemer: Father of the Steel Industry*, p. 87.

40. Carnegie lost confidence in Frick following the accident, writing that 'nothing I have ever had to meet in all my life, before or since, wounded me so deeply [as the Homestead incident]'. In 1894 he accepted Frick's resignation. K. H. Hillstrom and L. C. Hillstrom, *The Industrial Revolution in America*, Vol. 1: *Iron and Steel* (California: ABC-Clio, 2005), p. 87.

41. The sale was worth over 2 per cent of US GDP in 1901.

42. Elizabeth Bailey served as Dean of Carnegie Mellon University's Graduate School of Industrial Administration from 1983 to 1990. She was the first woman to receive a doctorate in economics from Princeton University, graduating in 1972.

43. Andrew Carnegie, *The 'Gospel of Wealth' and Other Writings* (New York: Penguin Books, 2006). 'The Gospel of Wealth' was first published as 'Wealth' in 1889 in *The North American Review*. The title was changed to 'The Gospel of Wealth' when published in London's *Pall Mall Gazette*.

44. Carnegie, *The 'Gospel of Wealth' and Other Writings*, p. 1.

45. A statistical measure of inequality is given by the Gini coefficient, where a value of one corresponds to maximum inequality and a value of zero corresponds to total equality. Most countries range between 0.25 and 0.6. Although global wealth inequality has been falling since the 1980s, in America the Gini coefficient has risen from 0.34 in the mid-1980s to 0.38 in the 2000s. The income of the super-rich has risen from twenty times the earnings of the lowest 90 per cent of America in 1980 to eighty times in 2006.

46. Carnegie, *The 'Gospel of Wealth' and Other Writings*, p. 10.

47. David Nasaw, *Andrew Carnegie* (New York: The Penguin Press, 2006), p. x.

48. Carnegie, *The 'Gospel of Wealth' and Other Writings*, p. 10.

49. Free university tuition for Scottish students is still the case today in spite of England charging tuition fees of up to £9,000 a year. *Securing a Sustainable Future for Higher Education*, 'The Browne Review', October 2010.

50. Ron Chernow, *Titan: The Life of John D. Rockefeller, Sr.* (New York: Random House, 2004), p. 313.

51. Chernow, *Titan*, p. 314.

52. Carnegie's vanity is exemplified by a story he tells in his autobiography. During the Civil War, he was sent to fix the railway lines between Baltimore and Annapolis Junction which had been cut by the Confederacy. En route, he noticed the telegraph wires had been pinned to the ground by wooden stakes and so called for the engine to stop. He rushed forward to release them, but as he did so the wires sprung up and hit him in the face, knocking him over and gashing his cheek. He writes: 'with the exception of one or two wounded a few days previously in passing through the streets of Baltimore, I can justly claim that I "shed my blood for my country" among the first of its defenders'. Yet, on closer examination, Carnegie's story 'doesn't bear scrutiny', writes David Nasaw, his biographer. Carnegie and his crew did not even repair that stretch of railway. 'As a little man and a foreigner, Carnegie needed to establish his credentials as man and patriot.' Andrew Carnegie, *Autobiography of Andrew Carnegie* (New York: Country Life Press, 1920), pp. 95–6. Nasaw, *Andrew Carnegie*, pp. 71–2.

53. The Frick Collection, housed in Henry Frick's mansion, is a jewel of a private collection. When I lived in New York, it was never crowded and so I went often to look at the works of the great masters in the unusually peaceful surroundings.

54. The word 'skyscraper' was first used to describe tall ships, but by the 1890s it was commonly being applied to buildings.

55. *Life*, 20 June 1901.

56. 'Streetscapes: The Flatiron Building; Suddenly, a Landmark Startles Again', *New York Times*, 21 July 1991.

57. Edward Steichen also photographed it, while in 1916 French cubist Albert Gleizes painted *Sur le flat iron*.

58. S. B. Landau and C. W. Condit, *Rise of the New York Skyscraper 1865–1913* (New Haven: Yale University Press, 1996), p. 304.

59. Alice Sparberg Alexiou, *The Flatiron* (New York: Thomas Dunne Books, 2010), p. 152.

60. Joseph Needham and Donald Wagner, *Science and Civilisation in China*, Vol. 5, Part 11 (Cambridge: Cambridge University Press, 2010), pp. 278–9.

61. The prodigious output of Chinese steel works was made possible by the invention of the blast furnace, sometime around the first century BC long before its appearance in Europe during the Middle Ages. The blast furnace allows molten metal simply to be tapped out of the furnace and so to be produced on a large scale. Its size is only restricted by the amount of ore and fuel available and the size of the workforce that can tend to it.

62. Donald Wagner, 'The cast iron lion of Cangzhou', *Needham Research Institute newsletter*, No. 10, June 1991, p. 3.

63. When Mao announced the Great Leap Forward in 1958, he called for a 19 per cent increase in steel production in that year alone. To meet the ambitious targets, backyard furnaces were encouraged in which peasants could melt down their pots and pans. By 1959 there were over half a million backyard furnaces in operation, but the wood used to smelt the iron led to widespread deforestation. Moreover, peasants now had little time to tend their crops or the agricultural tools with which to do so. Mao's misguided reforms resulted in a sudden drop in agricultural output and mass starvation.

64. R. M. Lala, *The Creation of Wealth: The Tatas from the 19th to the 21st Century* (New Delhi: Penguin Portfolio, 2006), pp. 31, 46.

65. Lala, *The Creation of Wealth*, p. 27.

66. Ibid.

67. Amartya Sen, *The Argumentative Indian* (London: Penguin Books, 2006), p. 338.

68. In the West, George and Richard Cadbury also believed that the welfare of their workers was integral to the success of their business. In 1878 they chose a rural site outside Birmingham, UK, to which they relocated their confectionary factory and built Bournville, a model village with comfortable and spacious houses, leisure facilities and good transport links. For the Tatas, it was their nationalist vision of India's development which drove their progressive business practices; for the Cadburys, it was their Quaker faith.

69. Lala, *The Creation of Wealth*, p. 29.

70. Ibid., p. 183.

71. In 'The Social Responsibility of Business is to Increase its Profits' (*New York Times Magazine*, 13 September 1970), Friedman writes that 'in his capacity as a corporate executive, the manager is the agent of the individuals who own the corporation or establish the eleemosynary institution, and his primary responsibility is to them'.

72. Sen, *The Argumentative Indian*, p. 335.

73. As a result of not pursuing profit at all costs, J. R. D. Tata, the chairman of Tata & Sons between 1938 and 1991, believed the company had 'sacrificed 100 per cent growth … But we wouldn't want it any other way.' Lala, *The Creation of Wealth*, p. 200.

CARBON

1. When a firework explodes, the energy released excites the electrons in atoms to a higher energy level. When the electrons fall back down to their original state they release this energy as light. The energy levels that an electron is allowed to move between are specific and different for each atom, so the particles of light, called photons, that are emitted by each atom also have a specific amount of energy. The colour of a photon depends on its energy, and so atoms of different elements will emit light of different colours.

2. Carbonisation is the process of turning organic matter into almost pure carbon. Firewood is carbonised by heating it in the absence of oxygen,

3. Wood is largely made from cellulose and lignin, molecules which are themselves made from atoms of carbon, hydrogen and oxygen. Paper is made from wood pulp and some early inks were made from lampblack, a type of soot produced by charring organic matter.

4. Carbon bonds with itself so readily because of the structure of its electron 'shells' and the properties of the bonds these electrons form. Its outermost shell contains four electrons, each of which it will readily share with another carbon atom and in doing so form a carbon-carbon bond. This process is repeated across multiple carbon atoms to form long chains or rings. Thanks to the positioning of a carbon-carbon bond relative to other bonds and the atom's nucleus, they are strong and stable, requiring a relatively large amount of energy to break.

5. In Primo Levi's *The Periodic Table*, he describes how a versatile carbon atom travels from limestone into tree then into a human brain, eventually ending up as a dot on piece of paper. Primo Levi, *The Periodic Table*. (London: Penguin Books, 2000, originally published in 1975), pp. 188-96.

6. Gunpowder was traditionally made from a careful balance of charcoal, saltpetre (potassium nitrate) and sulphur roughly in the ratio 10:75:15. Saltpetre is an oxidiser, providing oxygen to enhance the burning of the carbon fuel. According to

John Bates, writing in 1634: 'The SaltPeter is the Soule, the Sulphur the Life, and the Coales the Body of it.' Joseph Needham, *Science and Civilisation in China, Volume 5, Part 7* (Cambridge: Cambridge University Press, 1986), p. 111.

7. Graphene will be discussed later in the book, in *Silicon*.

8. John Julius Norwich, *A History of Venice* (London: Allen Lane, 1982), p. 631.

9. Here, again, context is crucial. Pure carbon does not easily combust and it is the carbon-hydrogen bond that is key to the great energy hidden in fossil fuels.

10. An exothermic reaction is one that releases energy from a system, usually in the form of heat.

11. When burnt to produce an equivalent amount of energy, coal produces almost one and a half times as much carbon dioxide as oil and almost twice as much carbon dioxide as natural gas.

12. The Four Great Inventions originated with sinologist and missionary Joseph Edkins. In 1859, he compared the Japanese to the Chinese, writing that 'they can boast of no remarkable inventions and discoveries, such as printing, paper-making, the mariner's compass, and the composition of gunpowder'. Joseph Edkins, *The Religious Condition of the Chinese* (London: Routledge, 1859), p. 2.

13. Joseph Needham, the English scientist, historian and sinologist, proposed his project *Science and Civilisation* to Cambridge University in 1948 as a single volume. It soon grew to over seven volumes, each of which are composed of multiple parts, spanning the breadth of science and technology throughout Chinese history. It is one of the most astounding works of scholarship of the last century. Needham died in 1995, but his work continues today at the Needham Research Institute in Cambridge, of which I am a Trustee.

14. Joseph Needham & Donald B. Wagner, *Science and Civilisation in China, Volume 5, Part 11* (Cambridge: Cambridge University Press, 2008), p. 369.

15. Thomas Malthus, *An Essay on the Principle of Population* (London: Routledge, 1996, originally published in 1798).

16. Kenneth Pomeranz deals with coal's role in China's relative decline in his book *The Great Divergence: China, Europe, and the Making of the Modern World Economy* (Princeton: Princeton University Press, 2000), p. 65. Differences in the geographical location of coal deposits, the quality of transport links, and factors facilitating the diffusion of knowledge between artisans, entrepreneurs, engineers and scientists, all may have contributed to the fact that an industrial revolution occurred in Europe and not China. Pomeranz claims that 'while overall skills, resource, and economic conditions in "China", taken as an abstract whole, may not have been much less conducive to a coal/steam revolution than those in "Europe" as a whole, the distribution of those endowments made the chances of such a revolution much dimmer.'

17. William Blake famously wrote of England's 'dark Satanic Mills' in his epic poem

Milton. Blake was a seminal figure in the Romantic movement which reacted against the Industrial Revolution by idealising nature.

18. Alexis de Tocqueville, *Journeys to England and Ireland* (New York: Anchor Books, 1968, edited by J. P. Meyer, originally published in 1835), p. 96.

19. Asa Briggs, the English historian, *Victorian Cities* (Berkeley: University of California Press, 1993), p. 317.

20. In 1845, Friedrich Engels published a damning study on the harmful effects of the Industrial Revolution in *The Condition of the Working Class in England* based on his experience of living in Manchester between 1842 and 1844.

21. The Select Committee on Accidents in Mines of 1835 estimated just over 2,000 mining deaths for the preceding twenty-five years, although incomplete records likely make the reality far higher.

22. Choke damp (carbon dioxide) and white damp (carbon monoxide) suffocated workers, while fire damp (methane) caused explosions.

23. An 1842 Report by a Parliamentary Royal Commission on the employment of women and children in mines led to the Mines and Collieries Bill in the same year.

24. In 1698, Thomas Savery patented an 'atmospheric pump' that could raise water to a height of several metres. By injecting steam into a chamber and then cooling it with a jet of cold water a partial vacuum was created that forced water upwards. The pressure was too low to be of much use for mining, but in 1712 Thomas Newcomen invented a more advanced first steam pump based on the same idea. Newcomen's pump used vacuum pressure to draw a piston attached to a rocker. This could then be attached to a series of pumps to draw water from a much greater height. Newcomen's engine could also be used to haul coal up the mineshaft. Humphry Davy's innovation was to use a fine mesh in his lamp to stop any flame propagating outside that could cause explosions.

25. Town gas is produced from coal when heated with steam and oxygen, breaking down into a noxious mix of methane, hydrogen, carbon monoxide and carbon dioxide.

26. The Gang of Four were a radical political faction of the Chinese Communist Party. The Gang was led by Jiang Qing, Mao Zedong's wife, and also consisted of Zhang Chunqiao, Yao Wenyuan and Wang Hingwen. After Mao's death they lost their power and went into hiding. BP's office and base during one early exploration project in the 1970s was a mock Tudor mansion in Shanghai, known as the Red House; it was rumoured to have been used by the Gang of Four as a bolthole shortly before they were arrested. Following their arrest, the Gang was blamed for the worst excesses of the Cultural Revolution and received sentences ranging from death (later commuted to life in prison) to twenty years in prison. Browne, *Beyond Business,* pp. 179-80.

27. The full extent of Beijing's pollution is more often seen during the winter and spring

because of the greater frequency of inversion layers. Normally air temperature decreases with increasing altitude, but during an inversion a warm layer of air above the ground traps cold air below, causing pollutants to build up. China's smogs are reminiscent of those that plagued London through much of the twentieth century. The 'Big Smoke' of 1952 was London's worst ever example of pollution by smog. Weather conditions trapped emissions from London's numerous coal power stations and coal-burning stoves in homes across the city – the particularly cold December weather leading families to burn much more coal than usual. The thick 'pea soup' smog reduced visibility to only eleven inches, causing trains and buses to crash. It is estimated that as many as 12,000 people may have died by inhaling the toxic air of that London winter.

28. The reality could be far higher: many mining accidents are covered up by local officials for fear of being held accountable by the ruling Communist Party or the exposure of their own illicit ties to companies involved.

29. China's increasing dependence on coal has caused the carbon dioxide it produces per unit of energy to rise in recent decades, while in most other nations it has decreased.

30. BBC News, 'China pollution "threatens growth"', 28 February 2011. www.bbc.co.uk

31. In June 2011 I heard Premier Wen Jiabao, originally a geologist, speak at the Royal Society in London where he was being awarded the King Charles II Medal. In his speech he outlined the important role that science and technology would play in this transition. 'The Path to China's Future', 27 June 2011, Royal Society, London.

32. Allan Nevins, the American historian, *Ford: The Times the Man, the Company* (New York: Charles Scribner's Sons, 1954), p. 48.

33. Henry Ford, *My Life and Work* (London: William Heinemann, 1922), p. 22.

34. Nevins, *Ford: The Times the Man, the Company*, p. 54.

35. Ibid., p. 17.

36. Ibid., p. 1.

37. Oil not only powered the new mode of transportation, but was integral to the roads on which automobiles travelled. Asphalt, made from the largest hydrocarbon molecules in crude oil, is used to pave roads with a smooth surface.

38. The high output from Ford's factory enabled him to reduce the price of the Model T to only $500 in 1913, around $12,000 today.

39. Petroleum is coined from Greek the words for rock (*petros*) and oil (*elaion*). The term petroleum first appears in Agricola's earlier work of 1546, *De Natura Fossilium* (New York: Geological Society of America, 1955. Translated from the first Latin edition of 1546 by Mark Chance Bandy and Jean A. Bandy), p. 61.

40. The British Library hold a first edition of *De re metallica*, previously owned by

Prince Henry, the son of King James I. The margins are scrawled with annotations, Prince Henry clearly taking an active interest in the workings of the mining industry.

41. Agricola, *De re metallica,* p. 583.

42. An account of the technological specifics of oil production can be found in Thomas C. Frick *Petroleum Production Handbook* (New York: McGraw-Hill, 1962).

43. Browne, *Beyond Business,* pp. 152-75.

44. The extraction of oil and gas is in general more costly than the extraction of coal, for three reasons. First, the extraction of previously unreachable reserves of oil and gas requires the development of new and more complex facilities which have to operate in increasingly inhospitable environments. In contrast, while coal mining technology has seen many advances, the basic principles remain the same. Second, the nature of the two extraction techniques is fundamentally different. Oil and gas can only begin to be extracted once a large initial investment has been completed. This raises the risk associated with a project and the cost of capital, as do the many failed exploration attempts prior to a successful find. Coal mining is incremental, though, and costs are more closely correlated with the rate of extraction. Third, gases and liquids require greater resources to transport and store than solids such as coal.

45. In 1856, Darcy published a relationship for the flow rate of water in sand filters in *Les Fontaines Publiques de la Ville de Dijon,* a report on the construction of the Dijon, France, municipal water system. He determined that the rate of flow of the fluid was proportional to pressure drop in the system, the cross-sectional area that the flow moves through and permeability of the medium. The rate of flow is also inversely proportional to the viscosity of the medium.

46. The injection of water and gas into a well is often referred to as improved oil recovery (IOR), rather than enhanced oil recovery. EOR and IOR were traditionally regarded as sequentially implemented secondary and tertiary processes after natural recovery has stopped (the primary process). Such a distinction is no longer made; often all artificial recovery processes are considered and partially implemented from the beginning of production from a well.

47. Heat is also used to extract oil from shale formations. Some shales would have produced oil and gas if they had been buried for long enough at high enough temperatures and pressures. These processes normally take millions of years, but can be accelerated by heating. Liquid oil from shale rock has been commercially produced in this way since the 1800s (and shale has been burnt as a solid fuel since prehistoric times), but it was always extracted as shale and then processed. Confusingly, there are other shales that will flow oil if they are fractured. This is the source of significant production in the USA.

48. In 1896, John J. McLauren wrote: 'A flame or a spark would not explode

Nitro-Glycerin readily, but the chap who struck it a hard rap might as well avoid trouble among his heirs by having had his will written and a cigar-box ordered to hold such fragments as his weeping relatives could pick from the surrounding district,' *Sketches in Crude Oil – Some Accidents and Incidents of the Petroleum Development in all parts of the Globe* (1896).

49. M. K. Hubbert, 'Nuclear energy and the Fossil Fuels', *Drilling and Production Practice*, American Petroleum Institute (1956).

50. Thomas Friedman, *Hot Flat and Crowded* (London: Allen Lane, 2008), p. 250.

51. Only a year before, an explosion on board the *Odyssey* on the opposite coast of Canada had resulted in 132,000 tonnes of oil being released into the sea. Fortunately, much of this had been carried away from the coastline, reducing the environmental impact.

52. I discuss the Texas City disaster in detail in *Beyond Business*, pp. 203–6.

53. J. D. Rockefeller quoted in Daniel Yergin's Pulitzer prize-winning history of the oil industry, *The Prize: The Epic Quest for Oil Money and Power* (New York: Free Press, 2009, originally published in 1991), p. 26.

54. Ibid., p. 81.

55. The fragments of Standard Oil still exist as well-known international oil companies today. The largest company formed, Standard Oil of New Jersey, making up almost half the total value of Standard Oil, is known today as Exxon. Standard Oil of New York later became Mobil; Standard Oil of California became Chevron; Standard Oil of Ohio became Sohio, later absorbed by BP; Standard Oil of Indiana became Amoco, which later merged with BP; Continental Oil became Conoco; Atlantic became part of ARCO and then BP.

56. Yergin, *The Prize*, p. 89.

57. Ida Tarbell, 'Character Study Part One', published in *McClure's Magazine*, July 1905.

58. On 29 September 1916, the *New York Times* reported on its front page that Standard Oil's booming stocks 'makes its head a billionaire'. Rockefeller's personal wealth dramatically increased from $300 million to just short of a billion when Standard Oil Trust was broken up. The shares he held in each of the resultant companies rocketed when trading opened on 1 December 1911, and continued to appreciate until they passed the billion dollar mark. When wealth is measured as a percentage of the economy, Rockefeller remains the wealthiest American ever to have lived.

59. Ron Chernow, *Titan* (New York: Random House, 1998), p. 314.

60. Ibid.

61. In the notorious 'loans for shares' scheme, Russia's oligarchs obtained many of the struggling state's assets for what was thought to be significantly less than their true value. See also note 64.

62. In 2010, Russian oligarch and philanthropist Vladimir Potanin announced that he

would give away all his fortune to good causes, rather than to his children. 'From oligarchy to philanthropy', *Financial Times*, 8 May 2011. www.ft.com

63. Potanin came from a well-off Soviet family and so was different from the other Russian oligarchs, many of whom were from poor Jewish backgrounds. Before his involvement in the loans-for-shares schemes he had been a Soviet official and then a banker.

64. This period of history is documented by Chrystia Freeland in *Sale of the Century* (London: Abacus, 2005). A recent trial between oligarchs Abramovich and Berezovsky has once again opened up this wild period of Russia history, detailing their extravagant lives and even details of a secretive payment by Abramovich for the release of two British hostages. Berezovsky sought to sue Abramovich for $5 billion, claiming that he was 'intimidated' into selling his shares in Russian oil giant Sibneft at a fraction of their value. Berezovsky managed to get the trial to be played out in London because he believed the law would be selectively applied against him in Russia. In the end, Berezovsky was unsuccessful: he lost the case.

65. In *The Prize*, Yergin writes: 'As the nineteenth century gave way to the twentieth, they looked to government to restore competition, control the abuses, and tame the economic and political power of the trusts, those vast and fearsome dragons that roam so freely across the country. And the fiercest and most feared of all the dragons was Standard Oil', (pp. 80–81).

66. The Seven Sisters were: Jersey (Exxon), Socony-Vacuum (Mobil), Standard of California (Chevron) and Texaco (the four Aramco partners) and then also Gulf, Royal Dutch/Shell and British Petroleum.

67. Although OPEC nations still produce about 40 per cent of the world's oil (and about 60 per cent of that traded internationally), as with any large and diverse group of nations, divisions and rivalry remain, and agreeing to production quotas works much better in theory than in practice. For example, Saudi Arabia, a member of OPEC, has a strong alliance with the US. Moreover, as global demand for oil surges, many oil-producing nations are struggling to keep up, leaving little spare capacity with which to control prices. Saudi Arabia has recently been the only nation with spare physical capacity to increase the amount of oil produced (potentially to lower prices). Coordinated cutbacks in production (potentially to increase prices) require significant cooperation.

68. Browne, *Beyond Business*, pp. 24–42.

69. Alyeska Pipeline Service Company, http://www.alyeska-pipe.com

70. BP had been kicked out of Kuwait when the government took complete control of the Kuwait Oil Company, part owned by BP, in 1975.

71. Browne, *Beyond Business*, pp. 58–75.

72. Carlos Andrés Pérez first came to power in 1974. Two years later he formed PDVSA during the wave of resource nationalisation of the 1970s. Oil prices were high and

many developing countries were taking the opportunity to use their oil wealth to accelerate development. But in the 1980s, when oil prices fell, so did state revenues. Pérez lost the 1979 presidential elections and spent the 1980s travelling the world, learning about the resource curse and alternative modes of economic growth. Returning as President in 1993, he sought to implement his new understanding of oil economics.

73. Browne, *Beyond Business.* p. 125.

74. David Ricardo. *On the Principles of Political Economy and Taxation* (London: John Murray, 1821. Originally published in 1817).

75. Yergin, *The Prize*, p. 120.

76. Algeria is a case a point. By leaving some business for the international oil companies to do, however small but with a promise that it might change in the future, they kept expertise in the country.

77. 'The devil's excrement', *The Economist,* 22 May 2003. www.economist.com

78. Yergin, *The Prize*, p. xv.

79. Thomas Friedman writes that, according to the first law of petropolitics, 'the higher the average global crude oil price rises, the more free speech, free press, free and fair elections, an independent judiciary, the rule of law, and independent political parties are eroded'. With high oil prices, there is no need for citizens to pay the government taxes, but as a consequence the government becomes less accountable to its citizens. Oil wealth is then used for populist spending, relieving the pressure for democratic change, but is also spent on more forceful oppression, such as extra policing and intelligence services. 'The first law of petropolitics', *Foreign Policy*, 15 April 2006.

80. Browne, *Beyond Business*, pp. 110–19.

81. Diamonds are formed deep in the earth's mantle where high pressures and temperatures of over 1,000 degrees Celsius force together small grains of pure carbon into a regular, almost unbreakable, lattice structure.

82. *A Rough Trade* (London: Global Witness, 1998).

83. Ibid., p. 2.

84. Following the publication of the Global Witness report, De Beers moved to guarantee that 100% of its diamonds are conflict free, and since its inception in 2003 has so certified them through the Kimberley Process.

85. *A Crude Awakening* (London: Global Witness, 1999).

86. In 2003, shortly after the EITI began, a similar 'transparency' initiative was created specifically for the selling of diamonds. The Kimberley Process requires compliant nations to sell diamonds in tamper-proof containers with a document which certifies they were not mined to fund war. In doing so it hopes to help restrict the profits from mineral wealth being used to fund conflict in the Congo, Sierra Leone and elsewhere. However, in 2011, Global Witness withdrew from the Kimberley

Process saying that 'the scheme has failed'. In particular they cited crimes against humanity committed in attempts to profit from diamond mining in Zimbabwe. The scheme only covers diamonds used by rebel movements or their allies, making it toothless against 'legitimate' regimes, such as Robert Mugabe's ZANU-PF.

87. In 2004, Angola began publishing data on its oil production and exports. Although the situation continues to improve, oil revenues are still not fully disclosed. A report published late in 2011 by Global Witness found a gap of $8.5 billion between the oil revenues reported by the Angolan Finance and Oil Ministries and what Sonangol (Angola's national oil company) recorded in its accounts.

88. As part of the implementation of the Dodd-Frank Act in the United States, the Securities and Exchange Commission ruled in August 2012 that all US-listed companies in the extractive industries must disclose details of payments made to foreign governments. At the time of writing, the European Union is on the path to a similar ruling.

89. John Browne, 'Europe must enforce oil sector transparency', *Financial Times*, 25 April 2012. www.ft.com

90. *Cannonball* was the first platform to be designed and constructed in Trinidad and Tobago, costing $250 million. www.bp.com

91. The energy equivalence of one barrel of oil in the form of natural gas would fill a volume of 160 cubic metres, almost the same volume as two Routemaster London double-decker buses.

92. At first the technology was very dangerous. The first commercial LNG plant, built in Cleveland Ohio, suffered an explosion in 1944 that killed 128 people. LNG did not take off until the 1960s when technology for shipping LNG was developed. Before then, LNG could only be used as an expensive means of gas storage.

93. In the middle of the nineteenth century, James Prescott Joule and William Thomson (later Lord Kelvin) studied the equivalence of heat and mechanical work. The cooling of gas in LNG trains works because of this equivalence. When gas is slowly forced from a container of small volume and high pressure into a container of larger volume and lower pressure, work is done by the gas on both sides ('work' is done when energy is used). But less work is done by the gas in the large-volume, low-pressure container than in the small-volume, high-pressure container. This difference in work (as long as the energy of the entire system remains constant) results in a temperature change in the gas.

94. Trinidad and Tobago experienced sixteen consecutive years of real GDP growth previous to 2008 as a result of economic reforms adopted in the early 1990s.

95. The price of natural gas may never converge to a single international price as, unlike oil, the costs of shipping natural gas in a liquefied state are likely to remain a large proportion of the overall price.

96. In the US, about a third of natural gas in consumed to produce electricity. Slightly less is used in industry and the rest for heating and cooking in the homes and in the commercial sector.

97. In 2011, Trinidad only exported 20 per cent of its LNG to the US, down from 70 per cent in 2007. It now increasingly relies on export markets in Europe, South America and Asia.

98. *Gasland* (2010), Josh Fox, New Video Group; and State of Colorado Oil and Gas Conservation Commission, Gasland Correction Document, 2010.

99. Following the 1973 oil embargo Nixon announced Project Independence declaring: 'Let us set as our national goal, in the spirit of Apollo, with the determination of the Manhattan Project, that by the end of this decade we will have developed the potential to meet our own energy needs without depending on any foreign energy source.' Yergin, *The Prize*, p. 599.

100. 'America's New Energy Reality', *New York Times*, 9 June 2012. www.nytimes.com

101. Thomas Malthus, *An Essay on the Principle of population* (Oxford: Oxford University Press, 2008), p. 62.

102. *The Limits of Growth* (1972), the Club of Rome, Earth Island Limited, London.

103. The Green Revolution was a series of advances in agricultural technology, such as new high-yield varieties of wheat, fertilisers, pesticides and irrigation infrastructure, that swept through the developing world from the 1940s to the 1970s.

104. The fourth IPCC assessment report, released in 2007, connected two stark and simple facts. First, the concentration of carbon dioxide, the predominant greenhouse gas, in the atmosphere is increasing because of human activity. Second, the temperature of the earth's surface is increasing. It stated that these two observations, were 'very likely' (meaning greater than 90 per cent probability) to be linked: increased greenhouse gas emissions are 'very likely' to be increasing the average global temperature.

105. Actual risk, as opposed to perceptions of risks, is simply the likelihood of an event happening multiplied by the damaging nature of the consequences of the event. Even if the chance of a catastrophic climatic event is very small, the consequences are so harmful that risk remains high.

106. The IPCC was set up in 1988 to examine the current state of scientific knowledge on climate change and the potential future environmental and socio-economic impacts.

107. 'IPCC Third Assessment Report, Working Group II: Impacts, Adaptation and Vulnerability', Technical summary, Group 1, p. 79. Climate change results from an imbalance in the Earth's emissions and absorption of carbon. While we understand the natural factors that alter this balance, the anthropogenic factors are not well understood. Greenhouse gases (GHGs) and aerosols have the greatest anthropogenic impact on the energy balance and have opposite effects:

GHGs lead to warming and aerosols (except black carbon) lead to cooling. The cooling effects of aerosols are much more uncertain than the warming effects of GHGs. Uncertainty about aerosols is so large that models do not exclude that their cooling compensates almost entirely the greenhouse warming. Even more uncertainty emerges when trying to model the future climate because of feedback effects. These result from changes to water vapour concentration, cloud and snow cover that occur as the Earth warms. Feedback effects are expected to multiply the effect of global warming, yet by how much is very uncertain.

108. According to Popper, no single experiment can ever wholly prove a scientific finding; rather additional experiments that are in agreement with the finding serve to *corroborate* it. Conversely, if a new, and reputable, observation is in disagreement then the scientific finding is said to be falsified. *The Logic of Scientific Discovery* (London: Routledge, 2002. Originally published in German in 1934).

109. Browne, *Beyond Business*, pp. 76–89.

110. *Addressing Climate Change*, 1997. www.bp.com

111. 'The Nobel Peace Prize 2007'. Nobelprize.org, *An Inconvenient Truth*, 2006.

112. Leaders agreed that they would work towards a new global treaty agreement in 2015, to be enforced in 2020. This would replace the Kyoto Protocol. The new agreement will remove the division between developed and developing countries, who are currently under no obligation to cut emissions. A commitment was also made to the creation of a new climate fund to help developing nations develop clean energy sources and adapt to any damaging consequences of climate change.

113. In *The Politics of Climate Change*, Giddens writes of the Giddens Paradox: 'Since the dangers posed by global warming aren't tangible, immediate or visible in the course of day-to-day life, however awesome they appear, many will sit on their hands and do nothing of a concrete nature about them. Yet waiting until they become visible and acute before being stirred to serious action will, by definition, be too late'. *The Politics of Climate Change* (Cambridge: Polity Press, 2011), p. 2.

114. The percentage of Americans and Western Europeans who view global warming as a threat fell by 10 per cent during the late 2000s financial crisis. Gallup Poll, 20 April 2011.

115. G. Hardin, 'The Tragedy of the Commons', *Science* 162 (3859): 1243–8 (1968).

116. China, and perhaps the US, are the only nations who could take unilateral action against climate change and see a noticeable and beneficial result as a change.

117. Quoted in Shogren, 'Kyoto Protocol', *AAPG Bulletin*, October 2004, V. 88, No. 9, pp. 1221–2.

118. This is known as the Jevons Paradox, first described by English economist William Stanley Jevons in *The Coal Question* (1865). Jevons was yet another reincarnation of Malthus, concerned that coal resources would soon run out because of

over-consumption. He warned against the idea that more economical use of coal would prevent this 'Malthusian Catastrophe'; using coal more efficiently would conversely increase our consumption of it.

119. The IPCC concluded that there is a potential of at least 2,000 gigatonnes of carbon dioxide of storage capacity in geological formations, about two orders of magnitude more than total worldwide carbon dioxide emissions each year.

120. In 2006, as the Miller oilfield in Scotland came to the end of its life, BP sought to implement just such an EOR project. The empty pipeline through which to pump carbon dioxide back into the field was already in place and would spark new life into the nearly depleted reservoir. Ultimately the project never happened because of delays in regulatory and financial support from the British government, on whom BP was dependent to make the project economically viable. As long as the costs of CCS remain lower than the revenue it can generate from enhanced oil recovery, there exists potential for CCS to be adopted as part of the upstream oil and gas industry. Carbon dioxide injection now accounts for more oil production in the US than any other enhanced oil recovery method.

121. EU member states have committed to cut emissions to 20 per cent below 1990 levels by 2020. The EU Emissions Trading System is the first and largest international carbon-trading scheme in the world. Under the scheme, factories and power plants have a cap on their yearly emissions, but within this cap they are granted emissions allowances which can be traded.

GOLD

1. 'Eldorado Raft', c. AD 700–1600. The raft measures approximately 10 by 20 centimetres. It was cast in one single piece using the lost-wax technique in a clay mould. It is made from a mixture of gold, copper and silver, known as tumbaga, the low melting point of which makes the material even easier to work.

2. The El Dorado ritual was described by historian Juan Rodriguez Freyle in *Conquest and discovery of New Granada* (1636). The ritual at the lake had ceased before the first Spaniards arrived in Colombia, but the story was passed on from descendants of the tribe to Freyle. However, the exact root of the legend of El Dorado is disputed. The legend was not connected with the Muisca area until many years after the first expeditions that sought to find the city of gold, El Dorado.

3. In contrast to Xue, the villainous moon Chíya signified darkness, sorcery, instinct and the sinister side of human nature.

4. Before gold's modern denotation as Au in the eighteenth century, the chemical symbol for gold was a representation of the Sun. Even Au is the stem of aurora, the Latin for 'shining dawn'.

5. There are exceptions to our universal desire for gold. In the mid-nineteenth

century, Fijian islanders captured a chest of gold coins from a trading brig close to the islands. When Captain Cook arrived on Fiji, he found them skimming them across the water. Again, in Africa early traders happily swapped gold in favour of salt, which they considered far more useful.

6. Peter Bernstein, *The Power of Gold* (New York: John Wiley & Sons, 2000), p. 121.

7. Wama Poma, Andean chronicler. Michael Wood, *Conquistadors* (London: BBC Books, 2000), p. 134.

8. John Hemming, *The Search for El Dorado*, (London: Michael Joseph, 1978), p. 126.

9. One tribal leader, known as Delicola, hearing of the fates of other chieftains, created a story about a great empire of gold that served to heighten the delirium of Pizarro's men for gold and send them on further into the jungle.

10. In 1548, Pizarro was hanged outside Cuzco after rebelling against the New Laws for the government of the Americas; he believed them to be too liberal.

11. By 1700 the world stocks of gold and silver were five times those of 1492.

12. During the sixteenth century Europe suffered a period of immense inflation, known as the Price Revolution. These financial problems were largely the result of profound social and political changes that were sweeping through Europe long before Columbus's arrival in the Americas. The large influx of gold into Spain only exacerbated the problem. The value of a precious metal is not absolute. When gold becomes more plentiful, its value decreases, causing the price of other commodities, which are bought with gold, to soar. The greed of the conquistadors was the seed of their own destitution.

13. B. Hammer and J. K. Norskov, 'Why gold is the noblest metal of all', *Nature*, Vol. 376, 20 July 1995.

14. The Mold gold cape is a piece of ceremonial dress made of solid gold dating from the European Bronze Age. It was found near the Welsh town of Mold by workmen quarrying for stone in the nineteenth century. The sheet of gold from which it was created was beaten out of single gold ingot and then embellished in astoundingly intricate detail so as to mimic strings of beads and folds of cloth.

15. The nomadic tribes of Scythia, an area that spread across parts of Central Asia and Eastern Europe, produced ornate decorative objects in gold, silver, bronze and bone from the seventh to the fourth centuries BC. The Hermitage owns the most spectacular collection of Scythian gold artefacts, such as shields, combs and bowls depicting wild animals and fighting warriors.

16. Marshall puts the date somewhere between 18 and 20 January. However, one of his workers writes in his diary entry on 24 January: 'This day some kind of mettle was found in the tail race that looks like goalds, first discovered by James Martial, the boss of the mill.' John Walton Caughey, *California Gold Rush* (Berkeley: University of California Press, 1975), p. 2.

17. H. W. Brands, *The Age of Gold* (London: Arrow, 2006), p. 16.

18. Ibid.

19. Peter Browning, *Bright Gem of the Western Seas* (Lafayette, CA: Great West Books, 1991), p. 3.

20. On 29 May the *Californian* newspaper announced: 'The whole country from San Francisco to Los Angeles, and from the sea shore to the base of the Sierra Nevada, resounds to the sordid cry of gold! GOLD!! GOLD!!! while the field if left half planted, the house half built, and everything neglected but the manufacture of shovels and pick axes …' Caughey, *California Gold Rush*, p. 21.

21. In one particularly brutal attack in 1850, labelled the 'Bloody Island Massacre' after the incident, a United States Cavalry regiment killed more than fifty native Pomo in Lake County, California. The attack was retribution for the killing of two brutal and sadistic settlers by a group of Pomo, and led to a foundation being set up by a survivor's descendants to improve relations between the Pomo and other residents of California.

22. An early method of assaying the purity of a coin was to use a 'touchstone'. The edge of the coin was rubbed on the stone and then compared to a series of stones rubbed with gold of a known purity. However, touchstone sets were often inaccurate and could be manipulated by the owner of the set for his own benefit.

23. Non-metallic coins were used before the inventions of gold and silver coins, such as the mollusc shells, called cowries, that were used by the ancient Chinese. But these objects were only of use for small transactions and were not valued outside their own culture.

24. In the British Museum in London, where I was a Trustee between 1995 and 2005, they own a large collection of Lydian coinage, from misshapen electrum lumps to the intricately stamped pure gold coins of Croesus.

25. King Croesus was concerned about the growing power of the Persians and so consulted the oracle at Delphi on the matter. The oracle told him that if he made war on the Persians he would 'destroy a great empire'. The oracle was renowned for the accuracy of its prophecies, and this one was no exception; a great empire was indeed destroyed, but, unfortunately for Croesus, it was his own. Bernstein, *The Power of Gold*, p. 35.

26. A gold ducat weighs 3.5g and is 0.997 fine, giving it a full quota of 24 carats (essentially 100 per cent pure).

27. Frederic Lane and Reinhold Mueller, *Money and Banking in Medieval and Renaissance Venice*: Vol. 1 (Baltimore: Johns Hopkins University Press, 1985), p. xiii.

28. A circle of raised beads was designed to be aligned with the edge of the coin metal to reduce this practice, but, apart from the ducat, the alignment of this circle was often off centre.

29. In one trial in March 1393, a man called Leonardo Gradengio was suspected of

clipping in Alexandria. The Venetian consul in Alexandria investigated and found nearly three hundred clipped ducats in his strongbox. He fled but was tried in absentia in Venice, where he was found guilty and charged with the loss of his right hand, both eyes and banishment. Fifteen years later his wife pleaded for a pardon on the basis of his youth at the time of the offence, and it was given. But when he was caught in Venice in 1413, the sentence was carried out anyway.

30. Thomas Levenson, *Newton and the Counterfeiter* (Boston: Houghton Mifflin Harcourt, 2009), p. 59.

31. Newton is also celebrated on the British two-pound coin, which is engraved with a quotation from one of his letters to Robert Hooke: 'Standing on the shoulders of giants'.

32. Thomas Levenson, *Newton and the Counterfeiter* (Boston: Houghton Mifflin Harcourt, 2009), p. 112.

33. William Jennings Bryan, 'Cross of Gold Speech', 1896. http://historymatters.gmu.edu/d/5354/

34. Bernstein, *The Power of Gold*, p. 279.

35. http://goldsheetlinks.com/production.htm

36. Sebastião Salgado, *Workers* (London: Phaidon Press, 1993), p. 19.

37. Serra Pelada is due to be reopened by a large mining company in 2013, but operated by heavy machinery rather than by hand.

38. John Maynard Keynes, *Essays in Persuasion* (London: Macmillan, 1984, originally published in 1931), pp. 183–4.

39. Our obsession with the Egyptian gold in Tutankhamun's tomb, for example, led us to overlook historically and archaeologically more significant discoveries.

SILVER

1. An account of the conquistadors' search for the silver mountain can be found in Enrique de Gandía's *Historia crítica de los mitos de la conquista Americana* (Buenos Aires: 1929), pp. 145–96.

2. In Ayamá language: 'Pachacamac janac pachapac guaccaichan'. Waldemar Lindgren & J. G. Creveling, 'The Ores of Potosí, Bolivia', *Economic Geology*, May 1928, Vol. 23, No. 3, p. 234.

3. In Capoche's *Relación general de la Villa Imperial de Potosí*. Peter Bakewell, *Miners of the Red Mountain: Indian Labor in Potosí 1545–1650* (Albuquerque: University of New Mexico Press, 1984), p. 3.

4. Lewis Hanke, *The Imperial City of Potosí* (The Hague: Martinus Nijhoff, 1956), p. 28.

5. Ibid., p. 2.

6. On one occasion Don Quixote de la Mancha quotes his servant, Sancho Panza, as

using the saying.

7. Lewis Hanke, *The Imperial City of Potosí* (The Hague: Martinus Nijhoff, 1956), p. 30.

8. Bakewell, *Miners of the Red Mountain*, p. 45. Indeed, the name of the forced-labour system, the *mita*, came from the pre-Conquest days under Inca rule. For the yanaconas, the free-floaters in Inca society who were previously attached to nobles and military leaders, the new arrangement could even have been beneficial. In the early years, over 7,000 yanaconas worked at mining or ore smelting in Potosí, many of them serving Gonzalo Pizarro. After providing two marks of silver a week they were allowed to keep anything else they produced; some became quite rich.

9. As the scale of mining at Potosí increased, Spaniards scoured the countryside for new workers, drawing the *mita* from an ever-greater distance. Those from the lower regions couldn't survive the cold harsh climate of Potosí and many died en route.

10. Dominican monk Santo Tomas's 1550 report to the Council of the Indies. Hanke, *The Imperial City of Potosí*, p. 25.

11. Each year, Luis Capoche wrote, fifty miners died in hospital 'swallowed alive' by the 'wild beast' that was Potosí. Combined with those who died outright in the mines, it is probable that a few hundred miners died each year as a result of mining accidents. Bakewell, *Miners of the Red Mountain*, p. 145.

12. Hanke, *The Imperial City of Potosí*, p. 25.

13. Many of the Bohemian silver mines were abandoned because of flooding or the persistence of 'noxious air' in the shafts. These problems were eased by the development of drainage and ventilation technologies, but were not effectively solved until the invention of the coal-powered steam pumps in the eighteenth century.

14. The modern 'dollar' can be traced back to the 'thaler' from Joachimsthal.

15. Agricola, *De re metallica*, p. xxv.

16. Ibid., p. 5.

17. Ibid., p. 5.

18. Agricola wrote that the miner must understand 'Philosophy', 'Astronomy' and 'Arithmetical Science' if he is to find veins, determine their direction and whether they can be mined profitably. Philosophy in its medieval meaning encompassed the natural sciences, while astronomy, it was generally believed, enabled you to determine the direction of veins. Agricola, *De re metallica*, p. 4.

19. Ibid., p. 36. People flocked to the Freiberg mines, as they would to Joachimsthal 350 years later. The nearby village of Christiandorf, grew suddenly and haphazardly, becoming known as Sachstadt, 'the town of the Saxons'. Otto financed the building of new town walls and gave 3,000 marks of silver to the nearby monastery. He also purchased huge swathes of land and stored in his treasury more than 30,000 marks

of silver, later seized by the Bohemians in 1189.

20. Most of these silver sources were first discovered in the twelfth and thirteenth centuries as traders and colonisers searched central Europe for ore veins. The region had been left relatively untouched, as it lay beyond the boundaries of the Roman Empire. Later discoveries on the border of Bohemia and Moravia kept increasing Europe's silver supply. In 1298 a vast silver reserve was found in Bohemia, producing over 20 tonnes of silver a year for the first forty years. But as the flow of silver from Europe's mines dried up, silver continued to move out of Europe along trade routes to the East. When people realised silver was disappearing, they began to hoard what little they had and in doing so created the Great Bullion Famine of the fifteenth century.

21. Agricola, *De re metallica*, p. 217. Although Agricola accepted the general belief in demons and gnomes, he was, in all other respects, very sceptical regarding the supernatural. The divining rod was, at the time, widely used as a tool for finding veins. On considering the merits of the arguments for the divining rod's use, Agricola concludes that a miner, 'since we think he ought to be a good and serious man, should not make use of an enchanted twig'. Ibid., p. 41.

22. Ibid., p. 18.

23. During this period, the Delian League, a voluntary alliance of Greek states, was essentially turned into an Athenian Empire. Most members of the Athenian league donated to a central treasury, the funds of which were used to increase the size of the formidable Athenian fleet further and to finance the Athenian building programme, not least the magnificent Parthenon dedicated to Athena.

24. Xenophon, *A Discourse Upon Improving the Revenue of the State of Athens.*

25. The Spartans fought against Athens in the Peloponnesian War, beginning in 431 BC. To fund the war, producing armaments and paying soldiers, the minting of Owls accelerated on a huge scale. Heavy borrowings were made from the treasuries created on the Acropolis as a wartime security.

26. Alcibiades, the defecting Athenian general, had explained to the Spartans that the occupation would cut off the Athenians from their homes, farmland and the silver mines of Laurium. Some 20,000 slaves deserted the area, many from Laurium's silver mines, removing one of Athens' dominant sources of revenue.

27. An international ratio between the value of gold and silver emerged at around 15.5 to 1, but in the 1870s this ratio began suddenly to rise upwards. Gold became untied from silver and established itself as *the* dominant element of value. Just as sudden changes in the ratio of the price of oil to the price of gas can indicate a fundamental change in how we produce and consume these energy sources, the sudden change in the ratio of gold to silver was indicative of a shift in the international monetary system and of human perceptions of the relative value of these two metals.

28. When a photon of silver strikes photographic paper, an electron may be released.

This electron can then turn a positively charged silver ion into a neutral atom of silver. The silver atom is unstable, but if enough photons strike simultaneously then many silver ions are turned into silver atoms, forming a stable 'latent image' site. The overall latent image that is formed can then be developed later on to reveal the photograph.

29. The darkening of silver salts by the sun was first noted in 1614 by Angelo Sal: 'When you expose powdered silver nitrate to the sun, it turns as black as ink.' However it was unclear if this was due to the heat or the light emitted from the sun. Helmut Gernsheim, *The Origins of Photography* (London: Thames & Hudson, 1982), p. 19.

30. 'An account of a method of copying paintings upon glass, and of making profiles, by the agency of light upon nitrate of silver. Invented by T. Wedgwood, Esq. With observations by H. Davy', *Journals of the Royal Institution of Great Britain*, 1802, Vol. 1, pp. 170–74.

31. William Henry Fox Talbot, *The Pencil of Nature* (London: Longman, Brown, Green & Longmans, 1844), p. 6.

32. Fox Talbot, 'Some Account of the Art of Photogenic Drawing, or the Process by Which Natural Objects May Be Made to Delineate Themselves without the Aid of the Artist's Pencil', read to the Royal Society on 31 January 1839.

33. Daguerre had developed these methods in a partnership with Nicéphore Niépce. Niépce created the world's first permanent photograph in 1826 using the light sensitivity of bitumen. He was also the first to write down the concept of photography in 1829: 'heliography consisted of spontaneously reproducing the image received in the camera obscura by the action of light, with all the graduations from black to white'. And in 1858 Nicéphore's cousin Niépce de Saint-Victor observed the effect of uranium on photographic plates, forty years before Becquerel. He did not, however, understand the significance.

34. He discovered that a combination of silver nitrate, acetic acid and gallic acid (gallo nitrate) would create a latent image on the paper which could be developed later on by recoating the image with gallic acid. Today film photographs are developed using a 'reducing agent' which acts to turn silver ions into silver metal. This reaction happens more quickly around the silver atoms formed in the latent image, and so by controlling the developing time an image is produced.

35. 'Calo' from greek *kalos*, meaning beautiful, good and useful.

36. The spread of the calotype was hindered by restrictive patents enforced by Fox Talbot in contrast to the daguerrotype which, sponsored by Arago, was an object of national pride and so given to the world. To make others aware and convince them of the importance of his discovery, Fox Talbot published his photographic work, *The Pencil of Nature*, in six parts between 1844 and 1846. Photography was an alien concept at the time, and Fox Talbot had to explain to readers how the plates were

created 'by mere action of Light upon sensitive paper' and how 'groups of figures take no longer time to obtain than single figures as would require, since the camera depicts them all at once'. Fox Talbot, *The Pencil of Nature*, Plate XIV.

37. Eastman chose 'Kodak' because he wanted a name that was distinctive, that couldn't be misspelled easily and that was easy to pronounce in most languages.

38. Brian Coe, *George Eastman and the Early Photographers* (London: Priory Press, 1973), p. 56.

39. *History of Kodak: George Eastman.* www.kodak.com

40. The box was only 8 centimetres high and 16 centimetres long. The lens was chosen so that everything further than a metre away from the camera was in focus.

41. In *The Kodak Primer*, Eastman wrote: 'The principle of the Kodak system is the separation of the work that any person whomsoever can do in [making] a photograph, from the work that only an expert can do … We furnish anybody, man, woman, or child, who has sufficient intelligence to point a box straight and press a button … with an instrument which altogether removes from the practice of photography the necessity for exceptional facilities, or, in fact any special knowledge of the art.' Michael Frizot, *A New History of Photography* (Milan: Könemann, 1998), p. 238.

42. Coe, *George Eastman and the Early Photographers*, p. 67.

43. 'Saigon Execution 1968', Eddie Adams, Associated Press. Sebastião Salgado, *Migrations*, (New York: Aperture, 2000), p. 209. *'Auschwitz Children'*, Photographer Unknown. Getty. *Queen Elizabeth II: Portraits by Cecil Beaton* (London: V&A Publishing, 2011).

44. Frizot, *A New History of Photography*, p. 591.

45. 'The Acknowledged Master of the Moment', *Washington Post*, 5 August 2004. In 1952 Cartier-Bresson published a book of his photographs entitled *The Decisive Moment* (New York: Simon & Schuster, 1952).

46. 'Eulogy: General Nguyen Ngoc Loan', *Time*, 27 July 1998.

47. Salgado, *Migrations*. p. 14.

48. BP had earlier, and unsuccessfully, applied for the Block 65 concession.

49. Sarir was the only producing oilfield owned by BP and Bunker until the nationalisation in 1971.

50. In 1979, Bunker was sued by BP in the UK for development costs owed under their joint operating agreement. BP was awarded $15,575,823 and £8,922,060. Bunker did not pay these sums. Bunker was later declared bankrupt in 1988, after a US Court ordered him and his brother William Herbert Hunt to pay $130 million in back taxes, fines and interest to the Internal Revenue Service, following their attempt to corner the global silver market.

51. In an interview with financial journal *Barron's* in April 1974. Harry Hurt III, *Texas Rich* (New York: W.W. Norton & Company, 1981), p. 320.

52. Bunker had a passion for racehorse breeding. He was well known on the

international racing circuit, amassing horses as he would later amass silver.

53. Hurt, *Texas Rich*, p. 325.

54. Bunker had two brothers, William Herbert Hunt and Lamar Hunt, also an elder sister, Margaret. Herbert Hunt was the brighter of the brothers who, unlike Bunker, succeeded in gaining a degree in geology. Lamar was a sports entrepreneur, making millions from World Championship Tennis.

55. Jerome Smith, *Silver Profits in the Seventies* (Vancouver: ERC Publishing, 1972), p. 31. Smith also pointed to the growing industrial use of silver in electronics and the reliance of the photography industry, most notably Eastman Kodak, on silver halide films. Like gold, silver is a very good conductor of electricity, so much so that during the Second World War, the Manhattan Project borrowed 13,500 tonnes of silver (worth more than $300 million) from the US Treasury. Copper was in short supply during the war and so they wanted to use silver instead to build giant electromagnetic coils used in the enrichment of uranium.

56. Evidence to a congressional committee in 1980. Stephen Fay,. *The Great Silver Bubble* (London: Hodder & Stoughton, 1982), pp. 29–30.

57. The Hunts believed that the US government might try and expropriate their hoard of silver and so moved it abroad. Their concerns were not unjustified; in 1933, President Franklin Roosevelt had made the hoarding of gold coin, gold bullion and gold certificates illegal, requiring all but a small amount of gold to be delivered to the Treasury in return for a fixed price that undervalued gold.

58. Hurt, *Texas Rich*, p. 409. The Hunts later pursued a lawsuit against Comex, the trading market, for changing the rules in the middle of the 'game'.

59. If sold ignoring tax consequences.

60. Except for liquidation orders.

61. Hurt, *Texas Rich*, p. 410.

62. Ibid., p. 416.

63. All the Hunts' main assets (energy, real estate and precious metals) had grown rapidly during the seventies. Before the crash they had assets worth $12 to 14 billion. Even after the crash they still had assets of around $8 to 10 billion and even if the price of silver dropped to zero they would have $6 to 7 billion left. They were not ruined, merely less wealthy.

64. Ibid., p. 420.

65. The 'sweat of the sun' and the 'tears of the moon' were Incan names for gold and silver.

URANIUM

1. The hypocentre is the point on the ground above which a bomb is detonated. In Hiroshima this point was 680 metres above ground.

2. In this section, I focus on Hiroshima as the first instance of uranium's destructive use as a city-destroying bomb. Nagasaki is the only other city to feel the full force of an atomic weapon; it was the US's target for the second (plutonium) bomb dropped three days later on 9 August 1945.

3. In May 1974, NHK Hiroshima radio station called for listeners to send in their own drawings of the event. More than 2,000 drawings were collected. Another call in 2000 elicited a further 1,338 drawings. Now the collection consists of over 3,600 pictures drawn by 1,200 sufferers.

4. 'Weakly Writhing' by Tomomi Yamashina, who was sixteen at the time of the bomb and standing 3,600 metres from the hypocentre in front of the Hiroshima First Army Hospital. She was seventy-two at the time of drawing in 2002 after a call for more pictures. Hiroshima Peace Memorial Museum, *A-bomb Drawings by Survivors* (Hiroshima: The City of Hiroshima, 2007), p. 82.

5. George Shultz is an American economist, statesman and businessman. He served as the US Secretary of Labor from 1969 to 1970, as the US Secretary of the Treasury from 1972 to 1974, and as the US Secretary of State from 1982 to 1989.

6. The context of the Second World War gives insights into why the bomb was dropped. The detonation of the bomb and subsequent Japanese surrender brought the Second Word War to a sudden end. The atomic bomb created a certainty in an uncertain war. It was thought that the alternative, a US-led land invasion of Japan, would result in a far greater number of casualties: 'It was a question of saving hundreds of thousands of American lives,' US President Truman later told a TV audience. But those lives lost and those saved are incomparable; the event itself is an act of inhumanity regardless of the context. Lansing Lamont, *Day of Trinity* (London: Hutchinson, 1966), p. 303.

7. Although Pierre ultimately died after being struck by a horse-drawn carriage, Marie died from aplastic anaemia as a consequence of radiation exposure.

8. E. Rutherford, 'Nuclear Constitution of Atoms', *Proc. Roy. Soc.*, A97, 374, 1920.

9. James Chadwick, 'Possible Existence of a Neutron', *Nature*, p. 312 (27 February 1932). Following his discovery of the neutron, Chadwick also went to work on the Manhattan Project.

10. Otto Hahn led the team that split uranium using neutrons. He was awarded the 1944 Nobel Prize for Chemistry for the discovery of nuclear fission. Meitner should have been jointly awarded it.

11. Protons and neutrons are held together in the nucleus by the strong force, one of

four fundamental forces in nature. The other three are the familiar gravitational and electromagnetic forces, and the more obscure weak force.

12. Otto Frisch, *What Little I Remember* (Cambridge: Cambridge University Press, 1979), p. 116. Frisch was later among the first to hold the initial samples of U-235 produced in the Manhattan Project. 'I had the urge to take one, as a paperweight, I told myself. A piece of the first uranium-235 metal ever made. It would have been a wonderful memento, a talking point in times to come.' Tom Zoellner, *Uranium: War, Energy and the Rock that Shaped the World* (New York: Viking, 2009), p. 64.

13. Lise Meitner and Otto Frisch, 'Disintegration of Uranium by Neutrons: a New Type of Nuclear Reaction', *Nature*, Vol. 143, No. 3615, p. 239 (11 February 1939). Otto Frisch, 'Physical Evidence for the Division of Heavy Nuclei under Neutron Bombardment', *Nature*, Vol. 143, No. 3616, p. 276 (18 February 1939).

14. Only 64kg of uranium was inside the Little Boy atomic bomb. Less than a kilogram of this underwent nuclear fission and a mere 0.7g was directly converted into energy. To destroy a city you do not need much matter.

15. Isotopes are chemically the same, containing the same number of protons in the nucleus, but differing numbers of neutrons.

16. Only 0.7 per cent of naturally found uranium is of this type.

17. Richard Rhodes, *The Making of the Atomic Bomb* (New York: Touchstone, 1988), p. 292.

18. Many question why he brought so destructive a weapon to the attention of an authority that could make it a reality. This question is answered by Otto Frisch, who, having the same idea as Szilárd, presented his idea to the British government: 'I have often been asked why I didn't abandon the project there and then, saying nothing to anybody. Why start on a project which, if it was successful, would end with the production of a weapon of unparalleled violence, a weapon of mass destruction such as the world has never seen? The answer was very simple. We were at war, and the idea was reasonably obvious; very probably some German scientists had had the same idea and were working on it.' Rhodes, *The Making of the Atomic Bomb*, p. 325.

19. Charles W. Johnson and Charles O. Jackson, *City Behind a Fence* (Knoxville: The University of Tennessee Press, 1981), p. 43.

20. The other major method of enrichment at Oak Ridge was by electromagnetic separation. First electrons are removed from atoms of uranium so that they become positively charged. When directed into a magnetic field, these charged ions of uranium will follow a curved path, but the heavier uranium-238 ions will be deflected less than the uranium-235 ions and so the two isotopes can be separated.

21. Zoellner, *Uranium* Charles W. Johnson and Charles O. Jackson, *City Behind a Fence* (Knoxville: The University of Tennessee Press, 1981), p. 66.

22. Michael H. Studer et al., 'Lignin content in natural Populus variants affects sugar

release', *PNAS*, Vol. 108, No. 15, pp. 6300–305 (12 April 2011).

23. Harry Truman announcing the bombing of Hiroshima, 6 August 1945. Harry S. Truman Library, 'Army press notes', Box 4, Papers of Eben A. Ayers.

24. Ibid.

25. *Space Adventures*, March 1960, Charlton Comics.

26. Scott Zeman and Michael Amundson, *Atomic Culture, How we Learned to Stop Worrying and Love the Bomb* (Boulder CO: University Press of Colorado, 2004), p. 15.

27. This was the stern warning given by General Groves, director of the Manhattan Project, in the introduction to 'Dagwood Splits the Atom', a pamphlet produced in consultation with the US Atomic Energy Commission in 1949. *Dagwood Splits the Atom* (New York: Kings Features Syndicate, 1949).

28. *Dallas Morning News*, 12 August 1945, section 4, p. 8.

29. *Eagle*, 1 August 1952.

30. Calder Hall was the first commercial-scale nuclear power plant in the world, producing tens of megawatts of electricity for civilian use. Both the US and Soviet Union had previously generated small quantities of electricity from atomic energy.

31. 'Queen Switches on Nuclear Power', BBC, 17 October 1956. www.bbc.co.uk

32. Ibid.

33. R. F. Pocock, *Nuclear Power: Its development in the United Kingdom* (London: Institution of Nuclear Engineers, 1977), p. 25.

34. Peter Hennessy, *Cabinets and the bomb* (Oxford: Oxford University Press, 2007), p. 48.

35. The reactor design at Calder Hall was named PIPPA, or Pressurised Pile Producing Power and Plutonium. Plutonium is produced in reactors as a result of neutron absorption by uranium atoms. Weapons grade plutonium has a high concentration of the Pu-239 isotope. To get this, uranium fuel must only be left in the reactor for a short period of time and, as a result, less of the energy in the fuel is harnessed for electricity generation.

36. Michihiko Hachiya, *Hiroshima Diary, The Journal of a Japanese Physician August 6–September 30, 1945* (London: Victor Gollancz, 1955), p. 35.

37. Robert Socolow, 'Reflections on Fukushima: A time to mourn, to learn, and to teach', *Bulletin of the Atomic Scientists*, 21 March 2011. Socolow writes: 'Unless a large dread-to-risk ratio is assigned to choices such as whether to eat or not to eat, the experts' models of risk will not match the choices.'

38. Any deaths resulting from Fukushima are very unlikely to be statistically detectable. More dangerous than radiation could be carcinogenic chemicals scattered about by the earthquake and tsunami.

39. The week before I arrived in Japan, high radiation levels had been measured on a street in the Setagaya ward of Tokyo. The source turned out be old bottles of radium

stored in a cellar below the street, only discovered because of the now widespread radiation monitoring.

40. Soviet Leader Mikhail Gorbachev on Soviet Central TV later that year said that the Chernobyl accident was 'a cruel reminder that mankind is still trying to come to grips with the fantastic, powerful force which it has brought into being'.

41. Widespread anxiety and depression were prevalent in the population surrounding Chernobyl. This resulted from a fear of radiation, but also from their relocation to uncontaminated areas. Increased anxiety and stress may have had further adverse health effects through changes in lifestyle, such as diet, smoking and drinking habits.

42. Outside the immediate disaster area, lower level radiation in the surrounding nations of Belarus and Ukraine led to thousands more cases of thyroid cancer, but this is usually treatable. A possible increase in the incidence of leukaemia has been observed in those workers involved in the clean-up of Chernobyl. There has been no observable increase, above background levels, of other types of cancer. *Source and Effects of Ionising Radiation*, United Nations Scientific Committee on the Effects of Atomic Radiation, UNSCEAR 2008 Report to the General Assembly, Vol. II, 'Health effects due to radiation from the Chernobyl Accident', United Nations, New York, 2011.

43. These fears continue today. On 20 July 2012, a batch of radioactive blueberries, containing nine times the recommended radiation limit, were found at a market in Moscow, believed to have come from the contaminated Chernobyl fallout region.

44. Forty-six per cent favoured maintaining Japan's reliance on nuclear power at its current levels, while 44 per cent thought the use of nuclear power should be reduced. Only 8 per cent wanted Japan to increase its use of nuclear power (about the same percentage as in previous polls over the last twenty years). *Japanese Resilient, but See Economic Challenges Ahead*, Pew Research Center, June 2011.

45. 'Kan heaps pressure on atomic plant operator', *Financial Times*, 15 March 2011.

46. The Fukushima Nuclear Accident Independent Investigation Commission ruled that: 'The TEPCO Fukushima Nuclear Power Plant accident was the result of collusion between the government, the regulators and TEPCO, and the lack of governance by said parties.' The National Diet of Japan, 2012, p. 16.

47. *Japanese Wary of Nuclear Energy: Disaster 'Weakened' Nation*, Pew Research Center, June 2012.

48. For each energy source, the average number of fatalities per Gigawatt-year of energy produced (counting accidents of five or more fatalities from 1969 to 2000) are:

Coal (without China): 0.60

Coal (with China, from 1994 to 1999): 6.1

Oil: 0.90

Natural Gas: 0.11

LPG: 14.9

Hydro: 10.3

Nuclear: 0.048

Source: Nuclear Energy Agency, *Comparing Nuclear Accident Risks with Those from Other Energy Sources* (OECD, 2010).

49. Ian Buruma, *Wages of guilt* (London: Atlantic Books, 2009), p. 99.

50. *The Treaty on the Nonproliferation of Nuclear Weapons* (UN, 1968), Article VI. www.un.org

51. Excluding Iran, these nations are: the United States, Russia, the United Kingdom, France, China, India, Pakistan, North Korea and Israel. Between them they hold an estimated 22,400 nuclear weapons, over 95 per cent of which belong to Russia or the United States.

52. Back in 1965, before becoming President (and later Prime Minister), Bhutto said: 'If India builds the bomb, we will eat grass or leaves, even go hungry, but we will get one of our own. We have no other choice.' Gordon Corera, *Shopping for Bombs* (London: Hurst & Company, 2006), p. 9.

53. Ibid., p. 10.

54. Khan once told an interviewer: 'If I escort my wife to the plane when she's flying somewhere, the crew will take notice of who she is and she will receive VIP treatment from the moment she steps on the plane. As for me, I can't even stop by the roadside at a small hut to drink chai without someone paying for me. People go out of their way to show their love and respect for me.' Zoellner, *Uranium,* p. 118.

55. 'Mutual Deterrence' Speech by Secretary of Defense Robert McNamara, 1962.

56. The 'Dead Hand' system was most famously depicted as the Doomsday Machine in the 1964 film *Dr. Strangelove.*

TITANIUM

1. Titanium's high strength-to-weight ratio is a result of the structure in which its atoms are arranged. The atoms in the metal are arranged in alternating layers. Titanium atoms bond in a 'hexagonal close-packed' structure, in which the atoms in every second layer will lie directly above each other (ABABAB). Iron atoms bond in a 'cubic close-packed' structure, in which atoms in every third layer will lie directly above each other (ABCABCABC). The density of atoms in a hexagonal close-packed structure is much smaller than that in a cubic close-packed structure, so that titanium is far lighter than iron, but remains very strong. Titanium's resistance to corrosion is, surprisingly, a result of the element's high reactivity. Titanium is so reactive that it bonds with oxygen in the air forming a very thin layer

of titanium dioxide on the metal surface. It is this layer which protects the metal from corrosion and which quickly re-forms if scratched away.

2. Uranium was named after Uranus, the planet, which was discovered a few years earlier by William Herschel. Klaproth writes: 'Wherefore no name can be found for a new fossil [element] which indicates its peculiar and characteristic properties (in which position I find myself at present), I think it is best to choose such a denomination as means nothing of itself and thus can give no rise to any erroneous ideas. In consequence of this, as I did in the case of Uranium, I shall borrow the name for this metallic substance from mythology, and in particular from the Titans, the first sons of the earth. I therefore call this metallic genus TITANIUM.' Martin Heinrich Klaproth, *Analytical Essays Towards Promoting the Chemical Knowledge of Mineral Substances*, Vol. 1, p. 210 (1801).

3. An air-breathing jet takes in air from the environment, rather than a liquid oxidiser, to mix with fuel in the combustion chamber.

4. B. R. Rich and L. Janos, *Skunk Works* (Boston: Little Brown and Company, 1994), p. 193.

5. The high temperature is a result of kinetic heating due to the compression of gases around the aircraft. Aluminium alloys can cope with temperature up to 130 degrees centigrade, reached between Mach 2 and Mach 3, but for anything higher titanium alloys must be used. At these temperatures no off-the-shelf electronics would work and the whole system had to be designed from scratch.

6. According to Kirchhoff's law of thermal radiation good absorbers are also good emitters. By painting the aircraft black, more heat was radiated away, reducing the wing temperature by around 35 degrees centigrade.

7. Small weight savings can dramatically reduce the fuel needed to boost a space vehicle into outer space. Fuel tanks were often made from titanium alloys for their high strength-weight and long-term chemical compatibility. For Apollo 11, titanium and aluminium were used extensively in the Lunar Module.

8. Norman Polmar, *Cold War Submarines* (Virginia: Potomac Books, 2004), p. 136.

9. Ibid., p. 139. *K-162* was an 'interceptor' submarine. Using a titanium alloy hull enabled *K-162* submarines to have a fifth less mass and to move a tenth faster than if steel had been used. It could accelerate to a speed of 45 knots.

10. The metallic parts of the world's first implanted artificial heart in 2001 were made from titanium.

11. Around two-thirds of all titanium metal is consumed by the aerospace industry. The use of titanium alloys in commercial Boeing aircraft continues to rise. Between 1960 and 1995 titanium's percentage of the aircraft's empty weight rose from virtually nothing to around 9 per cent and that used in the engine's weight rose to over 30 per cent.

12. Titanium is most commonly bound in its ore with oxygen but, unlike iron, this

oxygen cannot be removed using carbon. Titanium is so reactive that it will also bond with the carbon atoms, forming useless titanium carbide. Kroll solved this problem using a two-step chemical process in which titanium is first chlorinated to produce titanium tetrachloride (titanium atoms with four chlorine atoms attached) and then mixed with molten magnesium, which produces titanium sponge metal and magnesium chloride. The sponge metal then has to be further processed to produce titanium ingots. The process is very expensive because each stage is energy- and capital-intensive. Moreover, the hardness of titanium metal makes it very costly to machine and much metal is wasted in the process. New methods of separation by electrolysis (similar to the way that aluminium is extracted from its ore) are currently under development, but none have yet been commercially successful.

13. Lance Phillips and David Barbano, 'The Influence of Fat Substitutes Based on Protein and Titanium Dioxide on the Sensory Properties of Low-fat Milks', *Journal of Dairy Science*, Vol. 80, No. 11, November 1997, pp. 2726–31.

14. In June 1946, the world's largest ilmenite deposit was found in the Lake Allard area of Quebec. Quebec Iron and Titanium Corporation formed in August 1948 between Kennecott Copper and the New Jersey Zinc Company. Titanium most commonly occurs in nature as either ilmenite (titanium iron oxide) or rutile (titanium oxide).

15. Newton had a working theory by January 1666, but didn't publish his 'New theory about light and colours' in *Philosophical Transactions* until 1672.

16. According to Benjamin Haydon, the nineteenth-century historical painter and writer, during an 'immortal dinner' on 28 December 1817 hosted by Haydon and attended by William Wordsworth, Charles Lamb, John Keats, and Keats's friend Thomas Monkhouse, Keats joked that Newton 'has destroyed all the poetry of the rainbow, by reducing it to the prismatic colours'. He then proposed a toast to 'Newton's health, and confusion to mathematics'. Based on this story, Richard Dawkins entitled his book on the relationship between science and the arts *Unweaving the Rainbow* (London: Penguin Books, 1998). Benjamin Robert Haydon, *The Autobiography and Memoirs of Benjamin Robert Haydon*: Vol. 1 (London: Peter Davies, 1926, edited by Tom Taylor), p. 269.

17. Bernard Cohen, *Cambridge Companion to Newton* (Cambridge: Cambridge University Press, 2002), p. 230.

18. The Sun emits electromagnetic radiation of different wavelengths (for visible light, a type of electromagnetic radiation, these correspond to different colours) in different intensities. As well as visible light, the Sun emits electromagnetic radiation of longer and shorter wavelengths that are not detectable by the eye. Our eyes have developed so that the range of wavelengths they can detect are those that the Sun emits with maximum intensity.

19. The Sun appears golden because the Earth's atmosphere acts like a filter, scattering

longer, bluer wavelengths of light (making the sky appear blue) and leaving behind the shorter yellow and red wavelengths.

20. Pilkington Glass produced the first commercial self-cleaning windows in 2001.

21. Deyong Wu and Mingce Long, 'Realizing Visible-Light-Induced Self-Cleaning Property of Cotton through Coating N-TiO2 Film and Loading AgI Particles', *ACS Appl. Mater. Interfaces*, 2011, 3 (12), pp. 4770–74.

22. These solar cells, called Grätzel cells after their inventor Michael Grätzel, are a type of 'dye-sensitized solar cell' (DSC). However, most solar cells are of another type, made from silicon, which will be discussed in the next chapter. Michael Grätzel and Brian O'Regan, 'A low-cost, high-efficiency solar cell based on dye-sensitized colloidal TiO2 films', *Nature*, Vol. 353, pp. 737–40 (24 October 1991).

23. Titanium's use in cutlery sets was revealed to me by a former British intelligence officer based in Moscow, who recounted his surprise at being asked to go out and purchase one for analysis.

SILICON

1. Vannoccio Biringuccio, *Pirotechnia* (Cambridge, MA: MIT Press, 1966), p. 126. Biringuccio was a contemporary of Georgius Agricola. In *De re metallica* Agricola writes: 'Recently Vannoccio Biringuccio of Sienna, a wise man experienced in many matters, wrote … on the subjects of the melting and smelting and alloying of metals … by reading his directions, I have refreshed my memory of these things which I saw in Italy.' He did rather more than refresh his memory, extensively copying (but also extending) sections of Biringuccio's work. Yet *De re metallica* and *Pirotechnia* are distinct texts. Agricola writes in great detail on mining practices, which are only briefly considered in *Pirotechnia*. Biringuccio instead focuses on the extraction of metals from ore and the fabrication of metallic objects. For example, he writes at length on the fabrication of guns and bells (he was employed later in life to cast arms and construct fortresses for, among others, the Venetian Republic). Agricola can be considered the 'father of the mining industry', while Biringuccio is the 'father of the foundry industry'. Agricola, *De re metallica*, p. xxvii.

2. Biringuccio, *Pirotechnia*, p. 126. The story of the discovery of glass, retold by both Agricola and Biringuccio, originated with Pliny the Elder (*Natural History*, c. AD 79).

3. The first glass objects were beads, which imitate gemstones, from Egypt in the third millennium BC; Near Eastern glass was developed around 1600 BC.

4. Biringuccio, *Pirotechnia*, pp. 126–7.

5. Viscosity is the friction between the molecules of a liquid.

6. Glass easily breaks because it does not have a rigid crystal structure.

7. 'Black elephant in *pastra vitrea* with deep blue eyes, tusks and legs'. Purchased from Claudio Gianolla, Antiquario, San Marco 2766, 30124 Venezia.

8. Cappellin Venini and Company was founded in 1921 and the following year its first glass objects were exhibited at the Venice Biennale. In 1925 the company split up and Martinuzzi became the director of the new Venini and Co.

9. The glassmaking industry was not the only one to be concentrated in a specific area; the move was part of a broader Venetian economic plan for consolidation.

10. Biringuccio, *Pirotechnia*, p. 130.

11. Agricola, *De re metallica*, p. 592.

12. Agricola, *De Natura Fossilium* (New York: Geological Society of America, 1955. Translated from the first Latin edition of 1546 by Mark Chance Bandy and Jean A. Bandy), p. 111.

13. Cristallo glass was first produced around 1450 by Angelo Beroviero. The presence of iron oxide in sand and plant ash, which was often used as a fluxing agent, gave poor-quality glass a blue or green colour. Adding the right amount of manganese oxide would largely take out the colour of the glass, but a weak yellow or grey tint would persist. Muranese glassworkers used very pure silica and purified the plant ash into a white salt to remove most of the iron oxide from the glassmaking ingredients. Only a small amount of manganese oxide then needed to be added, reducing the chance of unwanted yellow tints. The result was a product that resembled quartz rock crystal. Quartz is atomically very similar to glass, as both are made from oxygen and silicon in a roughly 2:1 ratio. But in quartz the atoms are arranged in a rigid lattice, while the atoms in glass are disordered and resemble a liquid.

14. To produce a clear reflection, a flat, thin and clear piece of glass is needed. The difficulty in producing this meant that for a long time polished metal mirrors produced a clearer image than those made from glass. In ancient Greece and Rome, a mixture of copper and tine, or sometimes bronze, was used to make small mirrors, usually for personal grooming.

15. Sabine Melchior-Bonnet, *The Mirror: A History* (New York: Routledge, 2001), pp. 16–17.

16. Vannoccio Biringuccio in Melchior-Bonnet, *The Mirror*, p. 20.

17. Mark Pendergrast, *Mirror, Mirror* (New York: Basic Books, 2003 [ebook]), location 1813/5102.

18. Ibid., location 1818/5102.

19. Melchior-Bonnet, *The Mirror*, p. 47.

20. Pendergrast, *Mirror, Mirror*, location 1844/5102.

21. In 1674 George Ravenscroft invented lead crystal glass by adding lead oxide to silica. The clear, heavy glass refracts light at a greater angle than normal glass, resulting in diamond-like sparkling when it is cut in facets. English lead crystal made Venetian

glasswork unfashionable. In 1674, Ambassador Alberti noted that the glassworkers of Murano residing in London '... are unemployed; they die of hunger or emigrate'. Patrick McCray, *Glassmaking in Renaissance Venice* (Aldershot: Ashgate, 1999), p. 163.

22. James Chance was one of few graduates to receive first-class honours at Cambridge to go into a career in business. Many continued in academia, while those leaving the University would go into the Church, the armed forces or a career in law. All were more highly regarded than business. Of those graduating in the top ten in mathematics between 1830 and 1860, only 3 per cent chose business as a career. The same was true at Cambridge when I studied there in the 1960s. A few weeks before I graduated in 1969, I saw Brian Pippard, one of my most distinguished professors, coming towards me along King's Parade. As he passed, he turned to his colleague and said: 'This is Browne. He is going to be a captain of industry. Isn't that amusing?' A career in business was generally regarded as a waste of potential for students at the University.

23. Chance had agreed to purchase the rights for Bontemps' manufacturing techniques in return for five-twelfths of the profits. The development of patent law and the ability to profitably transfer know-how was important to securing a competitive advantage and rewarding innovation. Bontemps was not so certain about the security of patents, writing to Lucas in 1844 that 'A patent in general is fit only for an inventor, who has no manufactury of his own, and who wishes to sell the use of his invention'.

24. *Lancet*, Vol. 1, 22 February 1845, p. 214 (London: John Churchill, 1845).

25. Charles Ryle Fay, *Palace of Industry, 1851: A Study of the Great Exhibition and Its Fruits* (Cambridge: Cambridge University Press, 1951), p. 16.

26. *The Times*, 2 May 1851, in Patrick Beaver, *The Crystal Palace* (Chichester: Phillimore, 1986), pp. 41–2.

27. Ibid.

28. *The Times*, in Beaver, *The Crystal Palace*, p. 37.

29. Bessemer invented new methods for the production of glass lenses and plate glass. He also designed a new reverberatory furnace, which was sold to Chance Glass.

30. Paxton built the glasshouse to house a giant Amazonia lily he had grown. In awe at the sheer size of the plant, which had leaves over a metre and half in diameter, he used the rib structure that supported the leaves as a design for his glasshouse.

31. Before Paxton submitted his proposal, the Building Committee planned to build a brick and iron building four times the length of St Paul's cathedral with a huge iron dome placed at the centre. Many considered it a monstrosity. Paxton's design for the Crystal Palace was as pragmatic as it was aesthetic. Glass was cheaper than brick and could be assembled far more quickly.

32. Toby Chance and Peter Williams, *Lighthouses: The Race to Illuminate the World*

(London: New Holland Publishers Ltd, 2008), p. 108.

33. In 1854 the Crystal Palace was moved, strut by strut, pane by pane, to a new site on Sydenham Hill in south London. But on 30 November 1936, a small fire in the staff lavatory at the Crystal Palace quickly swept through the building, fuelled by the wooden floors and walls. All that remains today are the bare foundations.

34. *Punch*, 2 November 1951. The magazine, which also coined the name 'Crystal Palace', continued: 'We shall be disappointed if the next generation of London children are not brought up like cucumbers under glass' (Chance and Williams, *Lighthouses*, p. 109).

35. Cylinder-splitting glass was mechanised at the end of the nineteenth century, producing sheets up to 13 metres in length and almost 2.5 metres wide. Machine-blown glass, using compressed gas, was not introduced to Great Britain by Pilkingtons until 1909.

36. Ribbon glass had been produced for decades prior to Pilkington's invention. Pilkington Glass had developed the process with Henry Ford early in the twentieth century, who was trying to further reduce the cost of his automobiles. However, this ribbon glass still required grinding and polishing.

37. This is the Shard, which opened in July 2012 with a laser show across London.

38. *The Music Lesson*, Hiroshi Sugimoto (1999), author's collection.

39. Jonathan Miller, *On Reflection: An Investigation of Artists' Use of Reflection Throughout the History of Art* (New Haven: Yale University Press, 1998), p. 124.

40. *Republic* (X, 596) in Melchior-Bonnet, *The Mirror*, p. 104.

41. Glass lenses, named for their resemblance to lentils or, in Latin, 'lenses', had been sold by spectacle makers since the middle of the fourteenth century, but the telescope was not invented until over a century later. In October 1608 the General Estates of The Hague received a petition from Hans Lippershey for a patent to build an instrument for 'seeing faraway things as though nearby'. Patent Application of Hans Lippershey, 2 October 1608. The Hague, Algemeen Rijksarchief, MSS 'Staten-Generaal', Vol. 33, f. 178v.

42. Michael Hoskin, *The Cambridge Concise History of Astronomy* (Cambridge: Cambridge University Press, 1999), p. 112.

43. Nicolaus Copernicus, *De revolutionibus*, 1543. Along with Galileo's observations, the accurate observations of Tycho Brahe, who set new standards for astronomy in the sixteenth century, and the new laws of planetary motion formulated by Johann Kepler, were crucial in formulating a plausible account of a Sun-centred model of the Universe.

44. Galileo's Principle of Inertia provided an explanation for this 'problem'. According to his principle, a body moving at a constant speed will continue moving at that speed unless it is disturbed. As the Earth, and everything on it, moves at a constant speed, no force is felt.

45. Hoskin, *The Cambridge Concise History of Astronomy*, p. 112.
46. Glass bends or 'refracts' different colours of light by a different amount. This is why Newton's glass prism produced a rainbow. For an image to be produced close to the lens, the light must be sharply refracted using a more curved, and so thicker, lens. But this lens will also increase the discrepancy in refraction between different colours of light, so that a blurred image is produced. Using a thinner, more gently curved lens produced a clearer image, but the much longer focal length meant that a much longer telescope tube was needed.
47. Herschel understood that telescopes collect light 'in proportion to their apertures, so that one with double the aperture will penetrate into space to double the distance of the other'. The brightness of a star decreases rapidly with its distance from Earth, and so a mirror which collects more light can see stars which are further away. Pendergrast, *Mirror, Mirror*, location 2024/5102.
48. Pendergrast, *Mirror, Mirror*, location 2024/5102.
49. Many of Herschel's telescopes were not made from silvered glass, but from speculum metal, a brittle and hard casting composed mainly of copper and tin.
50. *The Great Art of Light and Shadow* (1646), in Frank Kryza, *The Power of Light* (New York: McGraw-Hill, 2003), p. 36.
51. Many historians believe that it really is no more than a legend. Experiments have even been carried out to replicate the event; even with the intense Sicilian sun, the army's bronze reflectors would probably have resulted in little more than smoking and charring of the enemy's wooden ships.
52. He used the same mirror to melt gold ducats. Biringuccio, *Pirotechnia*, p. 387.
53. Leonardo's design used a mosaic of silvered glass pieces stuck to the bottom of the curved basin. In a note he wrote: 'With [this device] one can supply heat to any boiler in a dyeing factory. And with this a pool can be warmed up, because there will always be boiling water.' Kryza, *The Power of Light*, p. 57. See also the image from Leonardo's notebook reproduced in this volume.
54. Bessemer's son, in Bessemer, *An Autobiography*, p. 36.
55. Shuman's hot box trapped heat in the same way as greenhouses do. Glass traps heat because it is transparent to electromagnetic radiation of optical wavelengths (visible light) which is emitted from the sun with a high intensity, but it is opaque to electromagnetic radiation of longer wavelengths, such as the infrared radiation that is dominantly emitted from the ground. Light enters in through the glass and is absorbed by the ground. When the energy is re-emitted as infrared radiation, it cannot pass through the glass and so the heat becomes trapped.
56. Kryza, *The Power of Light*, p. 11.
57. A. E. Becquerel, 'Mémoire sur les effets électriques produits sous l'influence des rayons solaires', *Comptes Rendus*, 9, pp. 561–7 (1839).
58. Pearson and Fuller's work at Bell Labs was not originally aimed at creating a

photovoltaic device; instead they were attempting to produce a better silicon transistor.

59. Yergin, *The Quest*, p. 570.

60. Silicon has four electrons in its outer shell, like carbon, and so bonds together to form crystals, in the same way that carbon atoms bond to form diamonds. To knock an electron out of this crystal structure requires a lot of energy and so, as an electrical current consists of flowing electrons, pure silicon at best can carry an extremely small current. A better semiconductor can be created by 'doping' the crystal with atoms of other elements. Either these atoms add extra electrons to the crystal, or they act as a 'hole' into which electrons can fall, so that a larger current can flow. If the semiconductor has an excess of electrons it is called an N-type (N for negative) semiconductor; if it has an excess of holes it is called a P-type (P for positive) semiconductor. A solar cell is made from a layer of N-type silicon sandwiched with a layer of P-type silicon. An electric field is created across the two layers between the negative free electrons and positive free holes. When a photon is absorbed by the solar cell it breaks apart an electron-hole pair into a free electron and a free hole, which are then swept to opposite sides of the device by the electric field. This flowing current can be harnessed as electrical power.

61. Silicon is an important tool on both the renewable and non-renewable sides of the transition. Shale gas is released from rock formations by the injection of sand (silicon dioxide) and water at high pressure.

62. 'Vast Power of the Sun Is Tapped By Battery Using Sand Ingredient', *New York Times*, 26 April 1954.

63. IBM 11230, Initial Press Release, IBM Data Processing Division, 11 February 1965.

64. In 2002, an unprecedented survey was carried out over the *Thunder Horse* field, producing around 28 terabytes of data, a billion times greater than the memory of the IBM 1130. It took BP's High Performance Computing Center in Houston about a month to process this data; but only two years earlier the same job would have taken almost two years, too long to be of value for the project.

65. Inside vacuum tubes hot filaments emit electrons that are then subsequently attracted to a positive voltage plate. As the plate is cold, and so does not emit electrons, the current flows in only one direction. If a negative voltage plate is placed between the filament and the positive plate, electrons are deflected away and current does not flow. Turn this plate off and current will flow as before. By turning the central negative plate on and off the vacuum tube can be used as a switch. The vacuum tube will also act to amplify the electrical signal put into the central negative plate.

66. Shockley's interest in semiconductors began during the Second World War, when he developed technology for the detection of radio waves. During the war, Shockley also influenced the decision to drop an atomic bomb on Hiroshima and Nagasaki

when he helped to produce a report on the probable casualties if Japan was invaded.

67. Bardeen and Brattain used a conducting fluid to create an electric field at the surface that broke down the surface state and so enabled current to flow.

68. The name of the transistor comes from 'transfer resistor'. Soon after the invention, co-worker John Pierce was walking by Brattain's office and was called in. Asked about a possible name for the device he recalled: 'I thought right there at the time, if not, within hours, I thought vacuum tubes have trans*conductance*, transistors would have trans*resistance*.' 'There were resistors and inductors and other solid states, capacitors and *tors* seemed to occur in all sorts of electronic devices. From transresistance I coined *transistors*.' Shurkin, *Broken Genius* [ebook], location 1889/5785.

69. Transistors are made from a stack of three semiconductor layers, either NPN or PNP (see note 60, above, for an explanation of doped semiconductor layers). The bottom layer acts as a source of charge carriers, the top layer acts as a drain of charge carriers and the middle layer acts as a channel through which charge carriers can (sometimes) flow. When the source and drain are attached to a battery, current cannot flow through the semiconductor layers. For example, in a PNP transistor, electrons will only flow from the negative battery terminal into the P-type semiconductor; both the positive battery terminal and the other P-type semiconductor that join the circuit are positively charged and so the charges will repel and no current will flow.

If, however, you inject some electrons into the middle layer of the sandwich through a 'gate', the electrons will move into the P-type semiconductor and a small current will flow. This small current acts as a switch that allows a much larger current to flow through the channel amplifying the original signal. In this way transistors act as both a switch and an amplifier.

70. *Fortune*, March 1953, p. 129, in Joel Shurkin, *Broken Genius* (London: Macmillan, 2006), p. 120.

71. At the time germanium was more readily available at the necessary purity than silicon. The first silicon transistor was not invented until 1954, but soon proved its worth. Silicon transistors work at high temperatures, vital for the military applications in which the first semiconductor devices were used.

72. *Fortune*, March 1953, p. 128.

73. Shockley wrote: 'I experienced some frustration that my personal efforts, started more than eight years before, had not resulted in a significant inventive contribution of my own.' Driven by his anger at not having followed through his transistor idea, Shockley set about designing a new and better type of transistor. He accomplished this a few months later. The 'junction transistor' is the forerunner of virtually all transistors used today. Shurkin, *Broken Genius*, pp. 107–8.

74. Christophe Lécuyer, *Making Silicon Valley: Innovation and Growth of High Tech,*

1930–1970 (Cambridge, MA: MIT Press, 2006), p. 133.

75. In computer factories, hundreds of lab coat-clad women would attach individual wires by hand to the join the devices.

76. Fairchild had a problem with the fragility of their silicon transistors: it would take little more than a sharp pencil tap against their transistors to stop them working. Jean Hoerni solved the problem when he discovered that the oxide layer usually washed off the semiconductor would actually protect its surface. He proved the invention to his co-workers by spitting on the surface.

77. Leslie Berlin, *The Man Behind the Microchip* (Oxford: Oxford University Press, 2005), p. 108.

78. At first no one knew how to do this and many saw little potential in the idea; it required such a step change in production that the first integrated circuits would be very expensive. Only the military would pay a premium for small improvements in weight and reliability. Yet Noyce saw the potential for his idea to revolutionise the computing industry and continued to support research into the development of the integrated chip at Fairchild. Gradually the importance of the idea became apparent. Noyce regarded his invention as a breakthrough in a process problem, rather than the discovery of any new science. Whenever asked when we would get his Nobel Prize for his achievement he always responded sardonically, 'They don't give Nobel Prizes for engineering or real work.' That situation I hope will change soon with the recent creation of the Queen Elizabeth Prize for Engineering, of which I am chairman. Noyce was never awarded a Nobel Prize, but would undoubtedly have shared the Prize given to Jack Kilby had he still been alive in 2000. www.qeprize. org. Berlin, *The Man Behind the Microchip*, p. 110.

Today integrated chips are produced by a method called photolithography. A thin silicon wafer is covered with a layer of insulating silicon dioxide and, on top of this, a layer of protective photosensitive material. When UV light is shone on to this material, the protective layer breaks apart and can be washed away. A mask is used so that the UV light only reaches parts of the chip where the circuit components are to be printed. After the protective layer has been washed away, chemicals are used to etch away the silicon dioxide in the same areas, revealing the silicon wafer beneath. The electrical properties of the silicon can now be altered as a first step to producing a transistor. For example, the silicon might be doped by adding other atoms of other elements to form one layer of a NPN or PNP junction (see note 60, above). This process is repeated to simultaneously build up all the components of the circuit. When all the components of the chip have been formed, a thin layer of metal is added over the top. The metal layer is then etched away so that the components are connected as desired. This is done using another photosensitive layer and another mask, this time in the shape of the connecting 'wires'. Complex circuits require many layers of components and metal 'wires'.

79. Berlin, *The Man Behind the Microchip*, p. 100.

80. Over the same period, the global silicon transistor business, focused in the US, grew from $32 million to around $90 million.

81. A resistor is used to restrict the flow of electrical current in a circuit, while capacitors are used as a store of electrical energy. Along with transistors, these are essential components in the building of logic gates.

82. Moore's law is not really a law, but, rather, an observation and a series of steps taken by the semiconductor industry. In fact, the continuing validity of Moore's law is partly self-fulfilling. The trend is one that those in the highly competitive computer industry recognise they must, at the very least, keep pace with, if they are to survive. In reality, the number of components on a chip double every eighteen months, rather than every year, as originally observed by Moore.

83. 'Cramming More Components Onto Integrated Circuits', *Electronics* magazine, 1965, in David C. Brock, *Understanding Moore's Law: Four Decades of Innovation* (Philadelphia: Chemical Heritage Press, 2006), p. 55.

84. Andrew Grove, *Only the Paranoid Survive* (London: HarperCollins, 1997).

85. Ibid., p. 30.

86. Information is encoded in particles of light called photons which are then sent down optical glass cables. Powerful lasers must be used to send light over long distances without the intensity dramatically decreasing. It was the invention of these lasers rather than glass fibres, which had existed for some time, that enabled the development of optical fibre communication systems in the 1970s.

87. Ray Kurzweil, *The Singularity Is Near: When Humans Transcend Biology* (London: Duckworth, 2005), p. 7.

88. Ari Shulman, 'Why minds are not like computers', *New Atlantis*, Winter 2009.

89. Gordon Moore, 'Moore's Law at 40', in Brock, *Understanding Moore's Law*, p. 6.

90. We create data at a rapidly increasing rate, whether on personal computers, in big data centres or scientific research institutions (such as CERN's Large Hadron Collider, which produces 15 million gigabytes of data every year). Copper struggles to handle such immense data flow; sufficient electrons cannot be moved fast enough over long distances. For example, big data centres are facing a 'performance bottleneck' because of their continued use of copper data connections. Currently, the processing cores and memory storage must sit side by side as copper cables can only transfer data over a short distance. Silicon photonics devices could enable these components to be separated so that all the processors, which produce most of the heat in a computer, can all be placed together. As a result, the power needed to cool the system would be dramatically reduced.

91. Named after futurist architect Buckminster Fuller for its resemblance to his geodesic dome designs.

92. Graphene's strength results from the powerful atomic bonds between carbon atoms and the flexibility of the bonds, which allow graphene to be stretched by up to a fifth of its normal size without being damaged.

93. In two-dimensional graphene, electrons can only move in the horizontal plane; there is no vertical movement. This dramatically reduces the rate at which electrons scatter off each other. The flow of electrons across graphene's surface is like the flow of cars down a motorway. The large number of electrons that can move at high speed, so-called 'ballistic conduction', results in graphene's extremely high electrical and thermal conductivity.

94. Interview with Novoselov, 19 March 2012.

95. K. S. Novoselov and A. K. Geim et al., 'Electric Field Effect in Atomically Thin Carbon Films', *Science*, 306, p. 666 (2004).

96. 'Physics Nobel Honors Work on Ultra-Thin Carbon', *New York Times*, 5 October 2010. www.nytimes.com. 'The Nobel Prize in Physics 2010'. Nobelprize.org

97. An electric current is produced in graphene when it absorbs photons. If a way can be found to harness this current, then it could be used to produce solar cells.

98. This is the Gartner's 'hype cycle'. After the initial excitement of a new discovery, there is usually a 'trough of disillusionment', before a gently rising productive plateau emerges. www.gartner.com

99. Most touch screens are made from an electrical insulator, such as glass, coated with a thin layer of transparent conductor. Electric current flows across the conductor. The human body conducts electricity and so when you touch the screen some current is drawn from the screen at the point of contact, changing the current flowing across the screen. The change in current is then measured by sensors at the edge of the screen and interpreted by the computer. This is why touch screens do not work when gloves are worn, since they are electrical insulators. Most touch screens are made using tin indium oxide for the conducting layer, but it is both expensive and breaks easily. Graphene is thin, hard-wearing and very conductive and so could make cheaper, faster and more enduring touch screens.

Power, Progress and Destruction

1. Agricola, *De re metallica*, p. 18.

2. Lieutenant General Brehon Somervell, Commanding General of the Army Services of Supply (at the time Groves was a colonel). Leslie Groves, *Now It Can be Told* (New York: De Capa Press, 1962), p. 4.

Acknowledgements

1. 'Unpacking my library; A talk about book collecting.' Walter Benjamin. *Illuminations* (New York: Schocken Books, 1969). My thanks to Dario Michele Zorza for providing this reference.

BIBLIOGRAPHY

The Essence of Everything

Bragg, W. H., *Concerning the Nature of Things* (London: G. Bell & Sons Ltd, 1925)

Browne, John, *Beyond Business* (London: Weidenfeld & Nicolson, 2010)

Diamond, Jared, *Guns Germs and Steel* (London: Jonathan Cape, 1997)

Feynman, Richard, *The Meaning of It All* (London: Penguin Books, 1998)

IRON

Ashton, T. S., *The Industrial Revolution* (Oxford: Oxford University Press, 1968)

Batty, Peter, *The House of Krupp* (London: Secker & Warburg, 1966)

Bessemer, Henry, *Sir Henry Bessemer, F.R.S.: An Autobiography* (London: Office of Engineering, 1905)

Bismark, Otto von, *Blut und Eisen* (1862)

Bodsworth, C., *Sir Henry Bessemer: Father of the Steel Industry* (London: IOM Communications Ltd, 1998)

BP, 'Building the Big One', *Frontiers*, April 2005, www.bp.com

Carnegie, A., *Autobiography of Andrew Carnegie* (New York: Country Life Press, 1920)

Carnegie, A., *The 'Gospel of Wealth' and Other Writings* (New York: Penguin Books, 2006)

Gillingham, John, *Coal, Steel and the Rebirth of Europe, 1945–1955* (Cambridge: Cambridge University Press, 1991)

Harris, F. R., *Jamsetji Nusserwanji Tata* (Oxford: Oxford University Press, 1925)

Hennessy, Peter, *Never Again* (London: Jonathan Cape, 1992)

Hillstrom, K. H. and Hillstrom, L. C., *The Industrial Revolution in America*, Vol. 1: *Iron and Steel* (California: ABC-Clio, 2005)

Hogg, Ian V., *German Artillery of World War Two* (London: Greenhill Books, 1997)

Howard, Michael, *The Franco-Prussian War* (London: Routledge, 2000)

Jeans, W. T., *Creators of the Steel Age* (London: Chapman & Hall, 1884)

Kraus, Peter, *Carnegie* (New Jersey: Wiley, 2002)

Krause, P., *The Battle for Homestead* (Pittsburgh: University of Pittsburgh Press, 1992)

Lala, R. M., *The Creation of Wealth* (New Delhi: Penguin Portfolio, 2006)

Landau, S. B. and Condit, C. W., *Rise of the New York Skyscraper 1865–1913* (New Haven: Yale University Press, 1996)

Manchester, William. *The Arms of Krupp* (London: Michael Joseph, 1968)

Manvell, R. and Fraenkel, H. *Adolf Hitler, The Man and the Myth* (New York: Pinacle, 1973)

Mills, Charles, *Echoes of the Civil War: Key Documents of the Great Conflict* (BookSurge Publishing, 2002)

Mindell, David, *Iron Coffin: War, Technology and Experience Aboard the USS Monitor* (Baltimore: Johns Hopkins University Press, 2000)

Morris, C., *The Tycoons* (New York: Times Books, 2005)

Nasaw, David, *Andrew Carnegie* (New York: The Penguin Press, 2006)

Needham, J. and Wagner, Donald B., *Science and Civilisation in China*, Vol. 5, Part 11 (Cambridge: Cambridge University Press, 2010)

Parker, William, *Recollections of a Naval Officer* (Annapolis: Naval Institute Press, 1985, originally published in 1883)

Roberts, William, *Civil War Ironclads* (Baltimore: Johns Hopkins University Press, 2002)

Schuman, Robert, *The Schuman Declaration*, 9 May 1950

Sen, Amartya, *The Argumentative Indian* (London: Penguin Books, 2006)

Sparberg Alexiou, Alice, *The Flatiron* (New York: Thomas Dunne Books, 2010)

Wagner, Donald B., 'The cast iron lion of Cangzhou', *Needham Research Institute newsletter*, No. 10, June 1991, pp. 2–3

Wawro, Geoffrey, *The Austro-Prussian War* (Cambridge: Cambridge University Press, 2003)

Weeks, Mary Elvira, *Discovery of the Elements* (Kessinger Publishing, 2003; first published as a series of separate article in the *Journal of Chemical Education*, 1933)

CARBON

Agricola, Georgius, *De re metallica* (New York: Dover Publications, 1950, translated by Hoover, H. C. and Hoover, L.H., originally published in 1556)

Agricola, Georgius, *De Natura Fossilium* (New York: Geological Society of America, 1955, translated from the first Latin edition of 1546 by Mark Chance Bandy and Jean A. Bandy)

AOGHS, 'Shooters – A "Fracking" History', *The Petroleum Age*, American Oil and Gas Historical Society, 4 (3): 8–9.

Bamberg, J. H., *The History of the British Petroleum Company*, Vol. 1: *1901–1932*. (Cambridge: Cambridge University Press, 1982)

Bamberg, J. H., *The History of the British Petroleum Company*, Vol. 2: *1928–1954*. (Cambridge: Cambridge University Press, 1994)

Bamberg, James, *British Petroleum and Global Oil*, Vol. 3: *1950–1975* (Cambridge: Cambridge University Press, 2000)

Berger, Michael, *The Automobile in American History and Culture* (London: Greenwood Press, 2001)

Blundell S. J. and Blundell, K. M., *Thermal Physics* (Oxford: Oxford University Press, 2007)

Briggs, Asa, *Victorian Cities* (New York: Harper & Row, 1963)

Brown, G. O., 'Henry Darcy and the making of a law', *Water Resources Research*, Vol. 38, No. 7, 1106, 10.1029/2001WR000727, 2002

Browne, John, *Addressing Climate Change*, 1997. www.bp.com

Chernow, Ron, *Titan* (New York: Random House, 1998)

Crowther, James, *The Cavendish Laboratory* (London: Macmillan, 1974)

Club of Rome, *The Limits of Growth* (London: Earth Island Limited, 1972)

Cullen, W. D., 'The public inquiry into the *Piper Alpha* disaster' (London: HMSO, 1990)

Darcy, H., *Les Fontaines Publiques de la Ville de Dijon* (Dalmont: Paris, 1856)

DHOS, 'Deep Water: Report to the President', National Commission on the Deepwater Horizon Oil Spill and Offshore Drilling, January 2011

Edkins, Joseph, *The Religious Condition of the Chinese* (London: Routledge, 1859)

Engels, Friedrich, *The Condition of the Working Class in England* (Oxford: Basil Blackwell, translated and edited by W. O. Henderson and W. H. Chaloner, 1958)

Flinn, Michael and Stoker, David, *The History of the British Coal Industry*, Vol. 2: *1700–1830: The Industrial Revolution* (Oxford: Clarendon Press, 1984)

Forbes, Robert, *Studies in Early Petroleum Histories* (Leiden: E. J. Brill, 1958)

Ford, Henry, *My Life and Work* (London: William Heinemann, 1922)

Franklin, Benjamin, 'Of the stilling of waves by means of oil', *Philosophical Transactions* 64: 445–60, 448, 1774

Freeland, Chrystia, *Sale of the Century* (London: Abacus, 2005)

Frick, Thomas C., *Petroleum Production Handbook* (New York: McGraw-Hill, 1962)

Friedman, Thomas, *Hot, Flat and Crowded* (London: Allen Lane, 2008)

Giddens, Anthony, *The Politics of Climate Change* (Cambridge: Polity Press, 2011)

Global Witness, 'A Rough Trade' (London, 1998)

Global Witness, 'A Crude Awakening' (London, 1999)

Goodell, Jeff, *Big Coal* (New York: Houghton Mifflin Company, 2006)

Hardin, G., 'The Tragedy of the Commons' Science 162(3859): 1243–8, 1968

Hart, Matthew, *Diamond* (London: Fourth Estate, 2002)

Helm, Dieter, *The Economics and Politics of Climate Change* (Oxford: Oxford University Press, 2009)

Hubbert, M. K., 'Nuclear energy and the Fossil Fuels', *Drilling and Production Practice*, American Petroleum Institute, 1956

International Energy Agency, 'Cleaner Coal in China', OECD, 2009

International Energy Agency, 'World Energy Outlook', 2011

IPCC, 'Third Assessment Report' Intergovernmental Panel on Climate Change, 2001

IPCC, 'Fourth Assessment Report' Intergovernmental Panel on Climate Change, 2007

Jevons, W. S., *The Coal Question* (1865)

Jianjun, Tu, 'Coal Mining Safety: China's Achilles' Heel', *China Security*, Vol. 3, No. 2, pp. 36–53 (World Security Institute, 2007)

Kurlansky, Mark, *Salt: A World History* (London: Jonathan Cape, 2002)

Lane, Frederic, *Venice: A Maritime Republic* (Baltimore: Johns Hopkins University Press, 1973)

Levi, Primo, *The Periodic Table* (London: Penguin Books, 2000, originally published in 1975)

Levine, Steve, *The Power and the Glory* (New York: Random House, 2007)

McLaurin, John J., *Sketches in Crude Oil – Some Accidents and Incidents of the Petroleum Development in all parts of the Globe* (1896)

Malthus, Thomas, *An Essay on the Principle of Population* (London: Routledge, 1996, originally published in 1798)

Montgomery, C. T. and Smith, M. B., 'Hydraulic Fracturing, History of an Enduring Technology', *Journal of Petroleum Technology*, December 2010

More, Charles, *Understanding the Industrial Revolution* (London: Routledge, 2000)

Morris, Ian, *Why the West Rules for Now* (London: Profile Books, 2010)

Needham, Joseph, *Science and Civilisation in China*, Vol. 1 (Cambridge: Cambridge University Press, 1954)

Needham, Joseph, *Science and Civilisation in China*, Vol. 5, Part 7 (Cambridge: Cambridge University Press, 1986)

Needham, Joseph and Golas, Peter J., *Science and Civilisation in China*, Vol. 5, Part 13 (Cambridge: Cambridge University Press, 1999)

Nevins, Allan, *Ford: The Times the Man, the Company* (New York: Charles Scribner's Sons, 1954)

Norwich, John Julius, *A History of Venice* (London: Allen Lane, 1982)

Peebles, Malcolm W. H., *Evolution of the Gas Industry* (London: Macmillan Press, 1980)

Pomeranz, Kenneth, *The Great Divergence: China, Europe, and the Making of the Modern World Economy* (Princeton: Princeton University Press, 2000)

Popper, K., *The Logic of Scientific Discovery* (London: Routledge, 2002, originally published in German in 1934)

Rackley, S., *Carbon Capture and Storage* (Amsterdam: Elsevier, 2010)

Ricardo, D., *On the Principles of Political Economy and Taxation* (London: John Murray, 1821, originally published in 1817)

Roston, Eric, *The Carbon Age* (New York: Walker & Co., 2008)

Schlumberger, 'Prize Beneath the Salt', *Oilfield Review*, Autumn 2008

Schlumberger, 'Has the Time Come for EOR?' *Oilfield Review*, Winter 2010/2011

Shepherd, R. and Ball, J., 'Liquefied Natural Gas from Trinidad and Tobago: The Atlantic LNG Project', James A. Baker III Institute for Public Policy (Energy Forum, 2004)

Skinner, S. K. and Reilly, W. K., 'The *Exxon Valdez* Oil Spill: A Report to the President', 1989

Song, Ligang and Woo, Wing Thye, *China's Dilemma* (Anu E Press, Asia Pacific Press, Bookings Institution Press, Social Science Academic Press, 2008)

Tarbell, Ida, 'Character Study Part One', published in *McClure's Magazine*, July 1905

Tarbell, Ida, *History of the Standard Oil Company* (New York: Philips & Co. 1904)

Tocqueville, Alexis de, *Journeys to England and Ireland* (New York: Anchor Books, 1968, edited by J. P. Meyer, originally published in 1835)

UNFCC, 'Kyoto Protocol', United Nations Framework on Climate Change, 1997

Victor, David et al., *Natural Gas and Geopolitics from 1970 to 2040* (Cambridge: Cambridge University Press, 2006)

Watts, Stephen, *The People's Tycoon* (New York: Alfred A. Knopf, 2005)

Werrett, Simon, *Fireworks: Pyrotechnic Arts and Sciences in European History* (Chicago: University of Chicago Press, 2010)

Whiteshot, Charles Austin, *The Oil-well Driller: A History of the World's Greatest Enterprise, the Oil Industry* (West Virginia: The Acme Publishing Company, 1905)

Winchester, Simon, *Bomb, Book and Compass* (London: Penguin Books, 2009)

World Bank, 'The Cost of Pollution in China' (2007)

Yergin, Daniel, *The Prize* (New York: Free Press, 2009, originally published in 1991)

Yergin, Daniel, *The Quest* (London: Allen Lane, 2011)

Zoellner, Tom, *The Heartless Stone* (New York: Picador, 2006)

GOLD

Ball, Phillip, *The Elements: A Very Short Introduction* (Oxford: Oxford University Press, 2004)

Bernstein, Peter, *The Power of Gold* (New York: John Wiley & Sons, 2000)

Brands, H. W., *The Age of Gold* (London: Arrow, 2006)

Bray, Warwick, *The Gold of El Dorado* (London: Royal Academy, 1978)

Browning, Peter, *Bright Gem of the Western Seas* (Lafayette, CA: Great West Books, 1991)

Caughey, John Walton, *California Gold Rush* (Berkeley: University of California Press, 1975)

Davies, Glyn, *A History of Money* (Cardiff: University of Wales Press, 2002)

Eichengreen, Barry and Flandreau, Marc, *The Gold Standard in Theory and History* (London: Routledge, 1997)

Emmerich, André, *Sweat of the Sun and Tears of the Moon* (Seattle: University of Washington Press, 1965)

Hammer, B. and Norskov, J. K., 'Why gold is the noblest metal of all', *Nature*, Vol. 376, 20 July 1995

Hammond, Innes, *The Conquistadors* (London: Collins, 1986)

Hemming, John, *The Search for El Dorado* (London: Michael Joseph, 1978)

Holliday, J. S., *The World Rushed In* (London: Victor Gollancz, first published in 1981, this edition in 1983)

Keynes, J. M., *Essays in Persuasion* (London: Macmillan, 1984, originally published in 1931)

Labbé, Armand, *Colombia Before Columbus* (New York: Rizzoli, 1986)

Lane, Frederic and Mueller, Reinhold, *Money and Banking in Medieval and Renaissance Venice*, Vol. 1 (Baltimore: Johns Hopkins University Press, 1985)

Lapiner, Alan, *Pre-Colombian Art of South America* (New York: Harry N. Abrams, 1976)

Levenson, Thomas, *Newton and the Counterfeiter* (Boston: Houghton Mifflin Harcourt, 2009)

Marx, Jennifer, *The Magic of Gold* (New York: Doubleday & Co., 1978)

Museo del Oro, *El Dorado: The Gold of Ancient Colombia* (New York: Graphic Society, 1975)

Pemberton, John, *Conquistadors* (London: Futura, 2011)

Ramage, Andrew and Craddock, Paul, *King Croesus Gold* (London: British Museum Press, 2000)

Rawls, J. J. & Orsi, R. J., *Mining and Economic Development in Gold Rush California* (Berkeley: University of California Press, 1999)

Rouillard, Patrick, *Colombia* (Publisher unknown, in EJPB library)

Salgado, Sebastião, *Workers* (London: Phaidon Press, 2003)

Shaw, Ian and Nicholson, Paul, *The British Museum Dictionary of Ancient Egypt* (London: British Museum Press, 1995)

Sutherland, C. H., *Gold: Its Beauty, Power and Allure* (London: Thames & Hudson, 1959)

Sykes, Ernest, *Banking and Currency* (London: Butterworth & Co., 1932)

Stahl, Alan, *Zecca* (Baltimore: Johns Hopkins University Press, 2000)

Weatherford, Jack, *A History of Gold and Money* (London: Pierre Vilar, 1976)

Wood, Michael, *Conquistadors* (London: BBC Books, 2000)

SILVER

Bakewell, Peter, *Miners of the Red Mountain: Indian Labor in Potosí 1545–1650* (Albuquerque: University of New Mexico Press, 1984)

Cartier-Bresson, Henri, *The Decisive Moment* (New York: Simon & Schuster, 1952)

Coe, Brian, *George Eastman and the Early Photographers* (London: Priory Press, 1973)

Coe, Brian, *Kodak Cameras: The First 100 Years* (Hove: Hove Foto, 1988)

Davy H., 'An account of a method of copying paintings upon glass, and of making profiles, by the agency of light upon nitrate of silver. Invented by T. Wedgwood, Esq. With observations by H. Davy', *Journals of the Royal Institution of Great Britain*, 1, pp. 170–74, p. 172 (1802)

Fay, Stephen, *The Great Silver Bubble* (London: Hodder & Stoughton, 1982)

Ferry, Stephen, *I Am Rich Potosí* (New York: Monacelli Press, 1999)

Fox Talbot, William Henry, *The Pencil of Nature* (Chicago: KWS Publishers, 2011) (London: Longman, Brown, Green & Longmans, 1844)

Fox Talbot, W. H., 'Some Account of the Art of Photogenic Drawing, or the Process by Which Natural Objects May Be Made to Delineate Themselves without the Aid of the Artist's Pencil', Read to the Royal Society on 31 January 1839

Friedman, Milton, *Money Mischief* (New York: Harcourt Brace Jovanovich, 1992)

Frizot, Michael, *A New History of Photography* (Milan: Könemann, 1998)

Gandía, Enrique de, *Historia crítica de los mitos de la conquista Americana* (Buenos Aires, 1929)

Gernsheim, Helmut, *The Origins of Photography* (London: Thames & Hudson, 1982)

Guilbert, John, *The Geology of Ore Deposits* (New York: Freeman, 1986)

Habashi, Fathi, 'Niece De Saint-Victor and the Discovery of Radioactivity', *Bull. Hist. Chem.*, Vol. 26, No. 2, 2001

Hanke, Lewis, *The Imperial City of Potosí* (The Hague: Martinus Nijhoff, 1956)

Howgego, Christopher, *Ancient History from Coins* (London: Routledge, 1995)

Hurt, Harry, *Texas Rich* (New York: W.W. Norton & Company, 1981)

Kagan, Donald, *The Peloponnesian War* (London: Harper Perennial, 2005)

Kraay, Colin M., *Archaic and Classical Greek Coins* (London: Methuen & Co. Ltd, 1966)

Lindgren, Waldemar and Creveling, J. G., 'The Ores of Potosi, Bolivia', *Economic Geology*, Vol. XXIII, No. 3, pp. 233–62, May 1928

Nef, John, *Cambridge Economic History of Europe*, Vol. II: *Trade and Industry in the Middle Ages*, Chapter X: 'Mining and Metallurgy in Medieval Civilisation' (Cambridge: Cambridge University Press, 1987, edited by M. M. Postan and Edward Miller)

Salgado, Sebastião, *Migrations* (New York: Aperture, 2000)

Smith, Jerome, *Silver Profits in the Seventies* (Vancouver: ERC Publishing, 1972)

Spufford, Peter, *Money and Its Use in Medieval Europe* (Cambridge: Cambridge University Press, 1988)

Spufford, Peter, *Power and Profit: The Merchant in Medieval Europe* (London: Thames & Hudson, 2003)

V&A, *Queen Elizabeth II: Portraits by Cecil Beaton* (London: V&A Publishing, 2011)

Wade, Nicholas J., 'Accentuating the negative: Tom Wedgwood' (1771–1805), photography and perception, *Perception*, Vol. 34, pp. 513–20, 2005

Wells, Liz, *Photography: A Critical Introduction* (London: Routledge, 2002)

Xenophon, *A Discourse Upon Improving the Revenue of the State of Athens*

URANIUM

Baggott, Jim, *Atomic* (London: Icon Books, 2009)

Boyer, Paul, *By the Bomb's Early Light* (New York: Pantheon, 1985)

Buruma, Ian, *Wages of Guilt* (London: Atlantic Books, 2009)

Cambridge University Physics Society, *A Hundred Years and More of Cambridge Physics* (Cambridge University Physics Society, 1974)

Chadwick, James, 'Possible Existence of a Neutron', *Nature*, p. 312 (27 February 1932)

Corera, Gordon, *Shopping for Bombs* (London: Hurst & Company, 2006)

Cunningham, C., 'The Silver of Laurion', *Greece & Rome*, Second Series, Vol. 14, No. 2 (October 1967), pp. 145–56

Degroot, Gerald, *The Bomb* (London: Pimlico, 2005)

Frantz, D. and Collins, C., *The Nuclear Jihadist* (New York: Hachette, 2007)

Frisch, Otto, *What Little I Remember* (Cambridge: Cambridge University Press, 1979)

Frisch, Otto, 'Physical Evidence for the Division of Heavy Nuclei under Neutron Bombardment', *Nature*, Vol. 143, No. 3616, p. 276 (18 February 1939) doi: 10.1038/143276a0

Frisch, Otto and Meitner, Lise. 'Disintegration of Uranium by Neutrons: a New Type of Nuclear Reaction', *Nature*, Vol. 143, No. 3615, p. 239 (11 February 1939) doi: 10.1038/143239a0

Gowing, Margaret, *Britain and Atomic Energy 1939–1945* (London: Macmillan, 1965)

Hachiya, Michihiko, *Hiroshima Diary, The Journal of a Japanese Physician August 6–September 30, 1945* (London: Victor Gollancz, 1955)

Hales, Peter, *Atomic Spaces: Living on the Manhattan Project* (Urbana, IL: University of Illinois Press, 1997)

Hendee, William, R., 'Personal and public perceptions of radiation risks', *Radiographics*, November 1991

Hennessy, Peter, *Cabinets and the Bomb* (Oxford: Oxford University Press, 2007)

Hiroshima Peace Memorial Museum, *A-bomb Drawings by Survivors* (Hiroshima: The City of Hiroshima, 2007)

IAEA, *The Great East Japan Earthquake Expert Mission*, International Atomic Energy Agency, 2011

International Institute for Strategic Studies, *Nuclear Black Markets: Pakistan, A.Q. Khan and the rise of proliferation networks* (London: IISS, 2007)

Johnson, Charles and Jackson, Charles, *City Behind a Fence* (Knoxville: University of Tennessee Press, 1981)

Lamont, Lansing, *Day of Trinity* (London: Hutchinson, 1966)

Lifton, Robert Jay, *Death in Life* (London: Penguin Books, 1971)

McNamara, Robert, 'Mutual Deterrence', 1962

National Diet of Japan, *The Fukushima Nuclear Accident Independent Investigation Commission*, 2012

Norris, Robert S. and Kristensen, Hans M., 'Global nuclear weapons inventories', 1945–2010, *Bulletin of the Atomic Scientists*, 2010 66: 77 doi: 10.2968/066004008

Nuclear Energy Agency, *Comparing Nuclear Accident Risks with Those from Other Energy Sources* (OECD, 2010)

O'Neil, John, *Almighty Atom: The Real Story of Atomic Energy* (New York: Ives Washburn Inc., 1945)

ORNL, *Oak Ridge National Laboratory, The First 50 Years*. www.ornl.gov

Pocock, R. F., *Nuclear Power: Its Development in the United Kingdom* (London: Unwin Brothers, Institution of Nuclear Engineers, 1977)

Rhodes, Richard, *The Making of the Atomic Bomb* (New York: Touchstone, 1988)

Rutherford, E., 'Nuclear Constitution of Atoms', *Proc. Roy. Soc.*, A97, 374, 1920

Socolow, Robert, 'Reflections on Fukushima: A time to mourn, to learn, and to teach', *Bulletin of the Atomic Scientists*, 21 March 2011

Studera, Michael H. et al., 'Lignin content in natural Populus variants affects sugar release', *PNAS*, 12 April 2011, Vol. 108, No. 15, pp. 6300–305

UN, *The Treaty on the Nonproliferation of Nuclear Weapons* (1968)

UNSCEAR, *Source and Effects of Ionising Radiation*, United Nations Scientific Committee on the Effects of Atomic Radiation, 2008 Report to the General Assembly, Vol. II, 'Health effects due to radiation from the Chernobyl Accident' (New York: United Nations, 2011)

Young, J. et al., *Radiation and Public Perception*, Advances in Chemistry (Washington, DC: American Chemical Society, 1995)

Zeman, Scott and Amundson, Michael, *Atomic Culture, How We Learned to Stop Worrying and Love the Bomb* (Boulder, CO: University Press of Colorado, 2004)

Zoellner, Tom, *Uranium* (New York: Viking, 2009)

TITANIUM

Aldersey-Williams, Hugh, *Periodic Tales* (London: Viking, 2011)

Cohen, B., *Cambridge Companion to Newton* (Cambridge, Cambridge University Press, 2002)

Crickmore, Paul, *Lockheed Blackbird: Beyond the Secret Mission* (Oxford: Osprey, 2004)

Crickmore, Paul. *Lockheed SR-71: Operations in the Far East* (Oxford: Osprey, 2008)

Crickmore, Paul. *Lockheed SR-71: Operations in Europe and the Middle East* (Oxford: Osprey, 2009)

Emsley, John, *Nature's Building Blocks* (Oxford: Oxford University Press, 2001)

Graham, Richard, *SR-71 Revealed* (Osceola: Motorbooks International, 1996)

Grätzel, Michael and O'Regan, Brian, 'A low-cost, high-efficiency solar cell based on dye-sensitized colloidal TiO_2 films', *Nature*, Vol. 353, pp. 737–40, 24 October 1991

Haydon, Robert, *The Autobiography and Memoirs of Benjamin Robert Haydon*: Vol. 1 (London: Peter Davies, 1926, edited by Tom Taylor)

Klaproth, Martin, *Analytical Analytical Essays Towards Promoting the Chemical Knowledge of Mineral Substances*, Vol. 1 (1801)

Leyens, Christoph and Peters, Manfred, *Titanium and Titanium Alloys, Fundamentals and Applications* (Köln: Wiley-VCH, 2003)

Mills, A. A., 'Newton's Prism and his Experiments on the Spectrum', *Notes Rec. R. Soc. Lond.* 1981, 36, 13–36 doi: 10.1098/rsnr.1981.0002

Phillips, Lance and Barbano, David, 'The Influence of Fat Substitutes Based on Protein and Titanium Dioxide on the Sensory Properties of Low-fat Milks', *Journal of Dairy Science*, Vol. 80, No. 11, November 1997, pp. 2726–31

Polmar, Norman, *Cold War Submarines* (Virginia: Potomac Books, 2004)

Polmar, Norman, *Submarines of the Russian and Soviet Navies 1718–1990* (Annapolis: Naval Institute Press, 1991)

RAND, 'Titanium: Industrial Base, Price Trends, and Technology Initiatives' (RAND Corporation, 2009)

Rich, B. R. and Janos, L., *Skunk Works* (Boston: Little Brown & Company, 1994)

Winkler, Jochen, *Titanium dioxide* (Hannover: Vincentz Verlag, 2003)

Wu, Deyong and Long, Mingce, 'Realizing Visible-Light-Induced Self-Cleaning Property of Cotton through Coating $N-TiO_2$ Film and Loading AgI Particles', *ACS Appl. Mater. Interfaces*, 2011, 3 (12), pp. 4770–74

SILICON

Auerbach, Jeffrey, *Great Exhibition of 1851* (New Haven, CT: Yale University Press, 1999)

Beaver, Patrick, *The Crystal Palace* (Chichester: Phillimore, 1986)

Becquerel, A. E., 'Mémoire sur les effets électriques produits sous l'influence des rayons solaires', *Comptes Rendus*, 9: 561–7 (1839)

Berlin, Leslie, *The Man Behind the Microchip.* (Oxford: Oxford University Press, 2005)

Biringuccio, Vannoccio, *Pirotechnia* (Cambridge, MA: MIT Press, 1966, edited by Cyril Stanley Smith and Martha Teach Gnudi, originally published in 1539)

BP, 'Computing Colossus' *Frontiers* (BP, 2003)

Bricknell, David, *Float: Pilkingtons' Glass Revolution* (Lancaster: Crucible, 2009)

Brock, David, *Understanding Moore's Law: Four Decades of Innovation* (Philadelphia: Chemical Heritage Press, 2006),

Ceruzzi. Paul, *A History of Modern Computing* (Cambridge, MA: MIT Press, 1998)

Chance Brothers Limited, *Mirror for Chance* (1951)

Chance, Toby and Williams, Peter, *Lighthouses: The Race to Illuminate the World* (London: New Holland Publishers, 2008)

Copernicus, Nicolaus, *De revolutionibus* (1543)

Deboni, Franco, *Venini Glass* (Turin: Umberto Allemandi, 2003)

Drake, Stillman, *Galileo: Pioneer Scientist* (Toronto: University of Toronto Press, 1990)

Fay, Charles Ryle, *Palace of Industry, 1851: A Study of the Great Exhibition and Its Fruits* (Cambridge: Cambridge University Press, 1951)

Gable, Carl I., *Murano Magic: A Complete Guide to Venetian Glass, Its History and Artists* (Atglen, PA: Schiffer, 2004)

Gibbs-Smith, C. H., *The Great Exhibition of 1851* (London: HMSO, 1981)

Grove, Andrew, *Only the Paranoid Survive* (London: HarperCollins, 1997)

Hoskin, Michael, *The Cambridge Concise History of Astronomy* (Cambridge: Cambridge University Press, 1999)

Kämpfer, Fritz, *Glass: A World History* (London: Studio Vista, 1966)

Kryza, Frank, *The Power of Light* (New York: McGraw-Hill, 2003)

Kurzweil, Ray, *The Singularity Is near* (London: Duckworth, 2005)

Lancet, Vol. 1: 22 February 1845, p. 214 (London: John Churchill, 1845)

Lécuyer, Christophe, *Making Silicon Valley: Innovation and Growth of High Tech, 1930–1970* (Cambridge, MA: MIT Press, 2006)

McCray, Patrick, 'Glassmaking in Renaissance Italy: The Innovation of Venetian cristallo', *Journal of the Minerals, Metals and Materials Society*, Vol. 50, No. 5 (1998), pp. 14–19 doi: 10.1007/s11837-998-0024-0

McCray, Patrick, *Glassmaking in Renaissance Venice* (Aldershot: Ashgate, 1999)

Melchior-Bonnet, Sabine, *The Mirror: A History* (New York: Routledge, 2001)

Miller, Jonathan, *On Reflection: An Investigation of Artists' Use of Reflection Throughout the History of Art* (New Haven: Yale University Press, 1998)

Novoselov, K. S., 'The rise of graphene', *Nature Materials*, Vol. 6, March 2007

Novoselov, K. S. and Geim, A. K. et al., Electric Field Effect in Atomically Thin Carbon Films', *Science*, 306, 666 (2004)

Pendergrast, Mark, *Mirror, Mirror* (New York: Basic Books, 2003 [ebook])

Rasmussen, S. C., *How Glass Changed the World* (Heidelberg: Springer, 2012)

Ryan, Johnny, *A History of the Internet and the Digital Future* (London: Reaktion, 2010)

Shurkin, Joel, *Broken Genius* (London: Macmillan, 2006 and ebook)

Siffert and Krimmel, *Silicon, Evolution and Future of a Technology* (Heidelberg: Springer, 2004)

Tait, Hugh, *Five Thousand Years of Glass* (London: British Museum Press, 1991)

EPILOGUE & ACKNOWLEDGEMENTS

Benjamin, Walter, *Illuminations* (New York: Schocken Books, 1969)

Groves, Leslie, R., *Now It Can Be Told* (New York: De Capa Press, 1962)

ENERGY AND ORES

Statistical data has mostly been derived from the BP Statistical Review of World Energy and the IEA World Energy Outlook

INDEX

1 49 16 — 4600 — 50
1 8 20 5100 — 55
22 29 56
26 28 61 —